GAUGE FIELDS

CLASSIFICATION

AND

EQUATIONS OF MOTION

GAUGE FIELDS
CLASSIFICATION
AND
EQUATIONS OF MOTION

M. Carmeli, Kh. Huleihil and E. Leibowitz

World Scientific
Singapore • New Jersey • London • Hong Kong

Published by

World Scientific Publishing Co. Pte. Ltd.
5 Toh Tuck Link, Singapore 596224
USA office: 27 Warren Street, Suite 401-402, Hackensack, NJ 07601
UK office: 57 Shelton Street, Covent Garden, London WC2H 9HE

British Library Cataloguing-in-Publication Data
A catalogue record for this book is available from the British Library.

GAUGE FIELDS
Classification and Equations of Motion

ISBN-13 978-9971-5-0745-9
ISBN-10 9971-5-0745-5

Preface

Although gauge fields are essentially quantum, the properties of the "classical" Yang-Mills field equations are of great importance. In particular, exact solutions of these equations, such as monopoles, dyons, instantons, and merons, have proved to be an essential and indispensable part of the whole theory.

In this monograph we discuss some recently obtained new exact solutions of the Yang-Mills SU(2) gauge field equations. These solutions are subsequently classified, the classification being similar to that of Petrov one has in general relativity theory in classifying the gravitational field. To obtain a better understanding of the physical meaning of the Yang-Mills fields, we also discuss the motion of a particle in these fields, first in general and later on for the particular fields we deal with in this monograph. Of particular interest is the motion in the instanton field where the equations of motion are written down, for particular cases, in terms of the angular momentum. The Yang-Mills field equations are highly nonlinear, and therefore special methods are used to solve them.

Beer-Sheva, Israel
October, 1988

Moshe Carmeli
Khadra Huleihil
Elhanan Leibowitz

Acknowledgements

We wish to thank Prof. S. Ben-Abraham for his useful remarks and suggestions on the exact solutions. We are grateful to Prof. S. Malin for his interest and encouragement. Also, we thank Prof. D.A. Owen for his interest in this work, Prof. N. Rosen and Dr. A. Lonke for their useful remarks. We are very much indebted to the late Mrs. Sara Corrogosky for her assistance. We are particularly grateful to Ms. Y.M. Chong, Editor of World Scientific Publishing Co., for the cooperation we had during the course of printing the book. Thanks are also due to Mrs. Rifka Shaanan for her help. Last, but not least, we wish to express our gratitudes to our spouses for the unlimited personal encouragement, patience and support we received during the course of writing and publishing this monograph and without which it would have never been published.

Table of Contents

Notations and Conventions

1. Greek indices α, β, μ, ν, ρ, ... take the values 0, 1, 2 and 3. They denote space-time indices.
2. Latin indices from i, j, k to m, n take the values 1, 2 and 3. They denote isospin vector indices and space indices.
3. Latin indices from a, b, c to f, g take the values 0 and 1. They are dyad indices.
4. Latin indices with primes are dyad indices in the conjugate space.
5. Latin capital indices A, B, ..., F, G take the values 0 and 1. They are spinorial indices.
6. The signature of the metric tensor $g_{\mu\nu}$ is $(+---)$.
7. g is the coupling constant.
8. $\varepsilon^{\alpha\beta\gamma\delta}$ is the contravariant tensor density of Levi-Civita, and is anti-symmetric in all the indices, with $\varepsilon^{0123} = 1$.

9. ε^{ab}, $\varepsilon^{a'b'}$, ε_{ab} and $\varepsilon_{a'b'}$ are the Levi-Civita anti-symmetric metric spinors, and are given by $\begin{pmatrix} 0 & 1 \\ -1 & 0 \end{pmatrix}$.

10. Partial differentiation is denoted by $\partial_\mu \phi = \partial\phi/\partial x^\mu$.

11. Differentiation with respect to the proper time τ is denoted by a dot, $d\phi/d\tau = \dot\phi$, and with a space variable by $\partial\phi/\partial x = \phi'$.

12. The symbol ∇ is a covariant differentiation.

13. The differential operator $\mathscr{D}$ is given by $\mathscr{D} = \partial/\partial\theta + (i/\sin\theta)\partial/\partial\phi$, where θ and ϕ are spherical coordinates.

1
Introduction

In a historical paper, published in 1954, Yang and Mills [1] formulated non-Abelian gauge theory. Consequently, a new era of modern physics began.

The non-Abelian gauge theory, as first presented by Yang and Mills, was based on the symmetry of the isospin. The concept of the isospin (isotopic spin) was first introduced by Heisenberg in 1932 [2] to describe the two states of the nucleon, the proton and the neutron, as two states of the same particle. It was a reasonable assumption because the proton and neutron have nearly the same mass, and the interactions $p - p$, $n - n$ and $n - p$ are identical when the electromagnetic interaction is neglected. This brought up the notion of the conservation of the total isospin in those interactions [3, 4]. The conservation of the isospin is equivalent to the invariance of the interactions under the rotation of the isospin. As a result, the orientation of the isospin has no physical meaning. But, if one chooses what to call a proton and a neutron at one space-time point, one is subsequently not free to make further choices at other space-time points. This is not consistent with the localized concept, and thus one has to look for a way by which the concept of conservation will be satisfied in a localized manner.

The situation is analogous to that in electromagnetism, where the field is described by a wave function ψ. Then, a change in gauge, namely, a phase transformation $\psi \rightarrow \psi' = (\exp i\alpha)\psi$, is of no physical meaning, and the choice of the phase α is arbitrary. To preserve the invariance of the equations of motion a potential field A_μ, which transforms under the gauge transformation in the form

$$A'_\mu = A_\mu + \frac{1}{e}\frac{\partial \alpha}{\partial x^\mu},$$

should be introduced to counteract the variation of the phase α [5].

In analogy with the electromagnetic gauge, Yang and Mills defined isotopic gauge as an arbitrary way to choose the orientation of the isospin in a local manner. They assumed that the physical processes are invariant under the gauge transformation,

$$\psi \rightarrow \psi' = S^{-1}\psi,$$

where S is a rotation transformation which depends on space-time coordinates. They introduced a potential field $\mathbf{b}_\mu$ to counteract the variation of S. The interaction of $\mathbf{b}_\mu$ with the sources of the isospin was fixed, similarly to the interaction of the electromagnetic potential A_μ with the electric charge. Furthermore, in the absence of the external sources there is an interaction of $\mathbf{b}_\mu$ with itself. In this property, $\mathbf{b}_\mu$ is different from the electromagnetic potential A_μ which is described by linear equations. The nonlinear equations, satisfied by $\mathbf{b}_\mu$, are similar to the equations in general relativity theory in some sense [6].

For a long time it was not known whether any of the processes in nature could be described by a non-Abelian gauge theory. Furthermore, it was not known if the symmetry could be broken spontaneously, nor if the theory could be renormalizable [7]. These problems, of great importance in quantum theory, were solved after two developments. The first, in 1964, when Higgs [8] introduced fields which induced spontaneous gauge symmetry breakdown. The second, in 1971, when 't Hooft [9] proved that the gauge theory can be renormalized. After these two developments, it became possible to quantize the non-Abelian gauge theory. Another important development was made by Weinberg [10] and Salam [11]. They constructed an SU(2) × U(1) model which unified the weak and the electromagnetic interactions. These achievements brought the idea that it is possible to construct a more general gauge theory by means of which all the processes in nature can be described.

At the same time, there was a development in describing the gravitational field and the Einstein field equations by using the gauge theory. This started after Utiyama [6], followed by Kibble [12], where they used the Lorentz and the Poincaré groups as the gauge group for the gravitational field. This, in turn, brought up a new formulation of gravitation as a gauge theory by Carmeli [13] who used the SL(2, C) group that gives the spinorial gravitational field equations, which are equivalent to the Einstein field equations, as previously formulated by Newman and Penrose [14]. This raised the possibility of renormalization of the gravitation field by Martellini and Sodano [15].

The Yang-Mills field equations are highly nonlinear, and are therefore difficult to be solved exactly. In similar cases, one has to perform approximate calculations using the perturbation method, but it is clear that there are phenomena which cannot be treated appropriately using such methods. Furthermore, there may be no small parameter in terms of which to expand or, if there is, the expansion may be singular. Thus, in the last years, new approximation techniques, which depend on the classical solutions, have been used [16]. One such method is to take the classical solutions and to quantize them using the semiclassical Wentzel-Kramers-Brillouin (WKB) method [17–19]. Another method is to use the Born-Oppenheimer fashion, in which the first term is calculated classically, and quantum corrections are found in series expansion [20]. Thus, the classical solutions are important for the quantum calculations.

Nevertheless, for many years there was little activity in the classical aspects of the Yang-Mills theory. The first non-Abelian solution for the Yang-Mills field equations was found by Wu and Yang in 1968 [21]. This solution is pointlike, its potential behaves like 1/r everywhere, and it describes a magnetic monopole. At the same time, a number of important works were published [22–27]. But a great advance was made in 1974, when 't Hooft [28] and Polyakov [29] obtained their magnetic monopole solution with a Higgs triplet as an external source, which is nonsingular and has a finite energy. Subsequently, there was a great deal of interest in exact solutions of the Yang-Mills field equations, and many works were done in this area [30-66].

Exact solutions, which are even more interesting than the monopole, were found in the Euclidean space-time E^4. The first solution was the instanton, obtained by Belavin *et al.* [67]. This solution is nonsingular, it is self-dual, and has a topological charge $q = 1$. Moreover, exact solutions describing an arbitrary number of instantons, were found [68]. These solutions are important in quantum Yang-Mills theory, since they are considered as vacuum states with transitions between them. Solutions of instantons were also obtained by Atiyah and Ward [69].

A different solution, also in the Euclidean space-time, is the meron [70–72]. This solution describes a pointlike density with a topological charge $\frac{1}{2}$, it is singular and is not self-dual. Solutions for two merons were also found. These solutions describe transitions between different vacuum states having topological charges 0 and $\frac{1}{2}$. It is believed that the instanton is composed of two merons [73, 74]. More information on classical solutions can be found in the review article of Actor [75]. Recently, more exact solutions of the Yang-Mills field equations were obtained [76–87].

Although there is a great deal of interest in the exact solutions of the Yang-Mills field equations, these equations have not been solved in general, and every new solution will give more understanding to the subject. In this monograph, we review some of the exact solutions of the Yang-Mills field equations, their classification and the motion of a classical test particle in these fields [88–93].

To classify these solutions, use is made of some known methods:

(a) The spinorial method of Carmeli, by means of which the Yang-Mills fields are classified according to the distinct eigenvalues and eigenspinors describing these fields [94–100].

(b) The matrix method of Wang and Yang, which depends on the rank of the matrix A ($A = E + iH$, where E and H are the electric and the magnetic parts of the Yang-Mills fields, respectively) [101].

(c) The little-group method of Castillejo, Kugler and Roskies. In this method, the Yang-Mills fields are classified according to the little group of $SL(2, C) \times SU(2)$ which leaves those fields invariant [102].

(d) The Anandan and Tod method [103].

Using the classification schemes, Castillejo and Kugler obtained all the solutions of class G according to their classification scheme [104]. It is worth mentioning that the use of classification schemes in general relativity gave the possibility of obtaining all the solutions of type D (Schwarzschild- and Kerr-type metrics) by Kinnersley [105].

To obtain solutions, use is made of the Yang-Mills field equations written in the null-tetrad formalism. As a result, the symmetry of the problem is exhibited, thus permitting the application of group theory to solve these equations. The null-tetrad theory of the Yang-Mills field equations was developed by Carmeli, Charach and Kaye [106] and, independently, by Newman [107, 108]. Using this formalism, new solutions were obtained [60, 109].

Recently, the problem of motion of a test particle in the classical Yang-Mills fields has stimulated a great interest, and many papers investigating this topic were published [110–116]. In this monograph, we give the equations of motion for a particle moving in the Yang–Mills fields, and their solutions in particular cases. The first is the magnetic solution field [90, 91], and the second is the instanton field [92].

In Chapter 2, we present the SU(2) Yang-Mills field equations in their standard form and in the null-tetrad formalism [106]. Here, we will restrict the form of the Yang-Mills field equations to the flat space-time case. Using this form, we will present some new solutions for the Yang-Mills field equations. The section is also written in order to fix our notations and conventions.

In Chapter 3, we present three known classification schemes for the Yang-Mills fields, by means of which the fields are classified using different criteria [94–102]. In Chapter 4, a comparison of two static solutions, obtained by Carmeli and Morris using different methods [60, 62], is made. We show that the two solutions coincide under certain conditions [88]. In Chapter 5, we classify [89] some gauge fields [109] according to the Carmeli classification scheme.

In Chapter 6, we present some exact solutions of the classifical Yang-Mills field equations. These solutions are obtained using the null-tetrad version of the equations [106] for some Yang-Mills fields classes. Consequently, some conditions on the Yang-Mills fields and potentials were obtained. By means of these conditions the exact solutions were derived [90, 91, 93].

The motion of a test particle in the classical Yang-Mills fields are reviewed in Chapter 7. Here, the works of Wong [110] and of Drechsler and Rosenblum [113] are presented. We use the equations of motion to examine the motion of a particle in two cases of Yang-Mills fields. These cases are the magnetic solution field [91] and the familiar instanton field [92]. Chapter 8 is devoted to the summary and concluding remarks. In the Appendices, we give the derivation of some formulas appearing in the text.

2

The Yang-Mills Field Equations and the Null-Tetrad Method

2.1 *The Yang-Mills field equations*

The requirement of local invariance under the rotation of the isospin brought up the introduction of a potential field by Yang and Mills [1], related to the isospin, in analogy to the electromagnetic potential with respect to the electric charge. In fixing this potential the following questions are raised:

(1) What kind of field does one have to introduce to preserve the invariance?
(2) How does this field change under the gauge transformation?
(3) What is the interaction form between this field and the isospin?
(4) What is the Lagrangian of this field?
(5) What field equations does it satisfy?

To answer these questions, Yang and Mills defined a gauge transformation in the form [1]

$$\psi = S\psi', \qquad (2.1.1)$$

where ψ describes a field with isospin $\frac{1}{2}$ and S is a 2×2 unitary matrix with determinant unity. In analogy to electromagnetism, the differentiation of ψ should then be replaced by the covariant derivative

$$(\partial_\mu - igB_\mu)\psi, \qquad (2.1.2)$$

where $\partial_\mu = \partial/\partial x^\mu$, the index μ takes the values 0, 1, 2, 3, $x^0 = t$, c and $\hbar$ are taken as units, g is the coupling constant, and B_μ are four 2×2 Hermitian matrices. The matrices B_μ were introduced to counteract the variation in S. From the invariance requirement, the covariant differentiation should satisfy the following:

$$S(\partial_\mu - igB'_\mu)\psi' = (\partial_\mu - igB_\mu)\psi. \qquad (2.1.3)$$

From Eqs. (2.1.1) and (2.1.3), one obtains the gauge transformation of B_μ:

$$B'_\mu = S^{-1}B_\mu S + \frac{i}{g}S^{-1}\partial_\mu S. \qquad (2.1.4)$$

It will be noted that the second term in Eq. (2.1.4) is similar to the gradient of the phase in the electromagnetic gauge transformation.

In analogy to the electromagnetic field strength, one defines the Yang-Mills field strength:

$$F_{\mu\nu} = \partial_\nu B_\mu - \partial_\mu B_\nu + ig(B_\mu B_\nu - B_\nu B_\mu). \qquad (2.1.5)$$

It is easy to show that the gauge transformation of $F_{\mu\nu}$ is given by

$$F'_{\mu\nu} = S^{-1}F_{\mu\nu}S. \qquad (2.1.6)$$

Other different functions of B_μ do not give field strengths which transform simply as (2.1.6).

From the matrices B_μ and $F_{\mu\nu}$, one defines the gauge potential $\mathbf{b}_\mu$ and field $\mathbf{f}_{\mu\nu}$ by

$$B_\mu = \mathbf{b}_\mu \cdot \mathbf{T}, \tag{2.1.7}$$

$$F_{\mu\nu} = \mathbf{f}_{\mu\nu} \cdot \mathbf{T}, \tag{2.1.8}$$

where the matrices $\mathbf{T} = (T_1, T_2, T_3)$ are a representation of the SU(2) generators which satisfy

$$[T_i, T_j] = i\varepsilon_{ijk} T_k. \tag{2.1.9}$$

Here ε_{ijk} is the Levi-Civita completely skew-symmetric tensor in three dimensions with $\varepsilon_{123} = 1$, the Latin indices take the values 1, 2, 3 and

$$\mathbf{T} = \boldsymbol{\sigma}/2, \tag{2.1.10}$$

where $\boldsymbol{\sigma} = (\sigma_1, \sigma_2, \sigma_3)$ are the Pauli matrices. From Eq. (2.1.5) one sees that the gauge potential and field are related by

$$\mathbf{f}_{\mu\nu} = \partial_\nu \mathbf{b}_\mu - \partial_\mu \mathbf{b}_\nu + g\mathbf{b}_\nu \times \mathbf{b}_\mu. \tag{2.1.11}$$

$\mathbf{f}_{\mu\nu}$ is a vector in the isospin space, while $\mathbf{b}_\mu$ is not a vector because of its transformation form. The symbol $\mathbf{b}_\mu$ stands for the triplet $(b_{\mu_1}, b_{\mu_2}, b_{\mu_3})$, and thus it is called *isotriplet*. To obtain the interaction between ψ, which describes the isospin field, and the potential $\mathbf{b}_\mu$, one has to replace the gradient of ψ by the covariant differentiation

$$(\partial_\mu - ig\mathbf{b} \cdot \mathbf{T})\psi. \tag{2.1.12}$$

Now, having defined the gauge field and its interaction with the isospin, one has to write down the Lagrangian density, and to obtain the field equations. Similarly to electromagnetism, the Lagrangian density of the gauge fields was given in the following form:

$$-\frac{1}{16\pi} \mathbf{f}^{\mu\nu} \cdot \mathbf{f}_{\mu\nu}.$$

Then the Lagrangian density, which includes the isospin too, is given by

$$\mathscr{L} = -\frac{1}{16\pi} \mathbf{f}^{\mu\nu} \cdot \mathbf{f}_{\mu\nu} - \bar{\psi}\gamma^{\mu}(\partial_{\mu} - ig\mathbf{T} \cdot \mathbf{b}_{\mu})\psi - m\bar{\psi}\psi, \qquad (2.1.13)$$

where γ_{μ} are the Dirac matrices.

Variation of the Lagrangian with respect to $\mathbf{b}_{\mu}$ and ψ then leads to the field equations

$$\partial_{\nu}\mathbf{f}^{\mu\nu} + g\mathbf{b}_{\nu} \times \mathbf{f}^{\mu\nu} + 4\pi\mathbf{j}^{\mu} = 0, \qquad (2.1.14)$$

$$\gamma^{\mu}(\partial_{\mu} - ig\mathbf{T} \cdot \mathbf{b}_{\mu})\psi + m\psi = 0, \qquad (2.1.15)$$

where

$$\mathbf{j}^{\mu} = \tfrac{1}{2}ig\bar{\psi}\gamma^{\mu}\boldsymbol{\sigma}\psi. \qquad (2.1.16)$$

It is easy to show that the divergence of $\mathbf{j}^{\mu}$ does not vanish, but it is equal to

$$\partial_{\mu}\mathbf{j}^{\mu} = -g\mathbf{b}_{\mu} \times \mathbf{j}^{\mu}. \qquad (2.1.17)$$

Hence one defines

$$\mathbf{J}^{\mu} = 4\pi\mathbf{j}^{\mu} + g\mathbf{b}_{\nu} \times \mathbf{f}^{\mu\nu}, \qquad (2.1.18)$$

which yields the conservation law

$$\partial_{\mu}\mathbf{J}^{\mu} = 0. \qquad (2.1.19)$$

It is seen that $\mathbf{J}_{\mu}$, as $\mathbf{b}_{\mu}$, does not behave as a vector under the gauge transformation. The time-independent total isospin, invariant under the Lorentz transformation, is then given by

$$\mathbf{T} = \int \mathbf{J}_0 d^3x. \qquad (2.1.20)$$

Finally, one defines the energy-momentum tensor of the Yang-Mills field in the following form:

$$T_{\mu\nu} = \frac{1}{4\pi}(\tfrac{1}{4}g_{\mu\nu}\mathbf{f}_{\alpha\beta}\cdot\mathbf{f}^{\alpha\beta} - \mathbf{f}_{\mu\alpha}\cdot\mathbf{f}^{\alpha}{}_{\nu}). \tag{2.1.21}$$

Here $g_{\mu\nu}$ is the metric tensor, a diagonal matrix with the values $(+1, -1, -1, -1)$.

From Eq. (2.1.18), one sees that the isospin current J^{μ} arises from both the field ψ of isospin $\tfrac{1}{2}$ and from the field $\mathbf{b}_{\mu}$. This fact leads to nonlinear field equations for $\mathbf{b}_{\mu}$, even in the absence of the isospin field ψ. In this respect, the gauge potential $\mathbf{b}_{\mu}$ is different from the electromagnetic potential A_{μ}. Equations (2.1.11) and (2.1.14) determine the gauge field and potential, and are called the Yang-Mills field equations.

2.2 *The null-tetrad method*

In this section, we give a review of the Carmeli-Charach-Kaye null-tetrad formulation of the Yang-Mills field equations in flat space-time. For the more general equations in curved space-time, the reader is referred to Ref. 106. A similar approach was developed for the electromagnetic field by Janis and Newman [117], and later on by Wódkiewicz [118]. Also, the null-tetrad method was used in the quantization of the electromagnetic and the gravitational fields [119, 120].

The use of null tetrads, in particular in flat space-time which is important from the point of view of field theory, helps to find exact solutions. In the case of electromagnetism, this leads to explicit use of the concept of null or general fields, obtained from the classification of the electromagnetic field.

In flat space-time, one defines four linearly independent null vectors, two of them, l^{μ} and n^{μ}, are real, and the other two, m^{μ} and $\bar{m}^{\mu}$, are complex where $\bar{m}^{\mu}$ is the complex conjugate of m^{μ}. The four null vectors satisfy the following orthogonality relations:

$$l_{\mu}l^{\mu} = n_{\mu}n^{\mu} = m_{\mu}m^{\mu} = \bar{m}_{\mu}\bar{m}^{\mu} = m_{\mu}l^{\mu} = m_{\mu}n^{\mu}$$

$$= \bar{m}_{\mu}l^{\mu} = \bar{m}_{\mu}n^{\mu} = 0, \tag{2.2.1}$$

$$l_{\mu}n^{\mu} = -m_{\mu}\bar{m}^{\mu} = 1. \tag{2.2.2}$$

These vectors can then be written as elements of the matrix

$$\sigma_{ab'}^{\mu} = \begin{pmatrix} l^{\mu} & m^{\mu} \\ \bar{m}^{\mu} & n^{\mu} \end{pmatrix}, \tag{2.2.3}$$

where the dyad indices, a and b', take the values 0, 1 and $0'$, $1'$, respectively.

The tetrad components of the Yang-Mills field and potential are given by

$$f_{ab'cd'k} = \sigma_{ab'}^{\mu} \sigma_{cd'}^{\nu} f_{\mu\nu k}, \tag{2.2.4}$$

$$b_{ac'k} = \sigma_{ac'}^{\mu} b_{\mu k}, \tag{2.2.5}$$

where the index k takes the values 1, 2, 3 and denotes the vector components in the isospin space.

It is known that any real anti-symmetric tensor with tetrad components $f_{ab'cd'}$ can be decomposed in the form [13]

$$f_{ab'cd'} = \varepsilon_{ac} \bar{\phi}_{b'd'} + \phi_{ac} \varepsilon_{b'd'}, \tag{2.2.6}$$

where

$$\phi_{ac} = \phi_{ca} = \tfrac{1}{2} f_{ab'c}{}^{b'}, \tag{2.2.7}$$

and $\bar{\phi}_{b'd'}$ is the complex conjugate of ϕ_{bd}. Here ε_{ac} and $\varepsilon_{b'd'}$ (and ε^{ac}, $\varepsilon^{b'd'}$ which will be used later on), are the Levi-Civita anti-symmetric spinors given by

$$\varepsilon = \begin{pmatrix} 0 & 1 \\ -1 & 0 \end{pmatrix}. \tag{2.2.8}$$

By means of these spinors, the spinor indices can be raised and lowered.

Using the above decomposition for the Maxwell tensor, the Maxwell equations can be written by means of the null tetrads [13, 121] or equivalently in terms of the dyad components of the Maxwell symmetric spinor ϕ_{ac}.

In the gauge theory case, the Yang-Mills field is given by the triplet of anti-symmetric spinors $f_{\mu\nu k}$. Hence the Yang-Mills theory is described by a triplet of symmetric spinors χ_{ack} along with potentials $b_{ab'k}$. In the Maxwell case, it is enough to use the fields or the potentials to describe the system. However, in the non-Abelian Yang-Mills theory, one should use both of them because of the coupling between the fields and the potentials on the one hand, and between different components of each triplet, on the other hand.

In analogy to the decomposition of the Maxwell field (2.2.6), the Yang-Mills field is decomposed as follows:

$$f_{ab'cd'k} = \varepsilon_{ac}\bar{\chi}_{b'd'k} + \varepsilon_{b'd'}\chi_{ack}. \tag{2.2.9}$$

Here

$$\chi_{ack} = \chi_{cak} = \tfrac{1}{2}\varepsilon^{b'd'}f_{ab'cd'k}, \tag{2.2.10}$$

and $\bar{\chi}_{b'd'k}$ is the complex conjugate of χ_{bdk}.

Explicitly, χ_{abk} is given by

$$\chi_{abk} = \tfrac{1}{2}\varepsilon^{c'd'}[\partial_{bd'}b_{ac'k} - \partial_{ac'}b_{bd'k} - (B_{bd'})_a{}^f b_{fc'k} - b_{af'k}(B_{d'b}^\dagger)^{f'}{}_{c'}$$

$$+ (B_{ac'})_b{}^f b_{fd'k} + b_{bf'k}(B_{c'a}^\dagger)^{f'}{}_{d'} + g\varepsilon_{kmn}b_{ac'n}b_{bd'm}], \tag{2.2.11}$$

where

$$\partial_{ab'} = \sigma_{ab'}^\mu \partial_\mu, \tag{2.2.12}$$

$B_{ab'}$ are the matrices [13]

$$B_{00'} = \begin{pmatrix} \varepsilon & -\kappa \\ \pi & -\varepsilon \end{pmatrix}, \tag{2.2.13a}$$

$$B_{01'} = \begin{pmatrix} \beta & -\sigma \\ \mu & -\beta \end{pmatrix}, \tag{2.2.13b}$$

$$B_{10'} = \begin{pmatrix} \alpha & -\rho \\ \lambda & -\alpha \end{pmatrix}, \tag{2.2.13c}$$

$$B_{11'} = \begin{pmatrix} \gamma & -\tau \\ \nu & -\gamma \end{pmatrix}, \tag{2.2.13d}$$

and $B_{d'b}^\dagger$ is the Hermitian conjugate of $B_{ab'}$.

The twelve elements ε, κ, π, etc. of the matrices $B_{ab'}$ are functions that were first introduced by Newman and Penrose [14] and are called the spin coefficients. They are expressed by the null tetrad and their differentiations as follows:

$$\varepsilon = \tfrac{1}{2}(l_{\mu;\nu}n^{\mu}l^{\nu} - m_{\mu;\nu}\bar{m}^{\mu}l^{\nu}), \qquad \kappa = l_{\mu;\nu}m^{\mu}l^{\nu},$$

$$\pi = -n_{\mu;\nu}\bar{m}^{\mu}l^{\mu}, \qquad \beta = \tfrac{1}{2}(l_{\mu;\nu}n^{\mu}m^{\nu} - m_{\mu;\nu}\bar{m}^{\mu}l^{\nu}),$$

$$\sigma = l_{\mu;\nu}m^{\mu}m^{\nu}, \qquad \mu = -n_{\mu;\nu}\bar{m}^{\mu}m^{\nu},$$

$$\tag{2.2.14}$$

$$\alpha = \tfrac{1}{2}(l_{\mu;\nu}n^{\mu}\bar{m}^{\nu} - m_{\mu;\nu}\bar{m}^{\mu}\bar{m}^{\nu}), \qquad \rho = l_{\mu;\nu}m^{\mu}\bar{m}^{\nu},$$

$$\lambda = -n_{\mu;\nu}\bar{m}^{\mu}\bar{m}^{\nu}, \qquad \gamma = \tfrac{1}{2}(l_{\mu;\nu}n^{\mu}n^{\nu} - m_{\mu;\nu}\bar{m}^{\mu}n^{\nu}),$$

$$\tau = l_{\mu;\nu}m^{\mu}n^{\nu}, \qquad \nu = -n_{\mu;\nu}\bar{m}^{\mu}n^{\nu},$$

where ; denotes a covariant differentiation.

Similarly to the Maxwell equations which are written by means of the spinor ϕ_{ab}, the Yang-Mills field equations are expressed by the spinor χ_{abk} in the following form:

$$\varepsilon^{ag}\partial_{gc'}\chi_{abk} = \varepsilon^{ag}(B_{gc'})^{d}{}_{b}\chi_{adk} - \varepsilon^{fg}(B_{fc'})^{a}{}_{g}\chi_{abk}$$

$$- g\varepsilon_{kmn}\varepsilon^{ag}\chi_{abn}b_{gc'm} + 2\pi j_{bc'k}. \tag{2.2.15}$$

The dyad components of the total isospin current and the energy-momentum tensor are then given by

$$J^{ab'}{}_{k} = 4\pi j^{ab'}{}_{k} + g\varepsilon_{kmn}b_{cd'm}f^{ab'cd'}{}_{n}, \tag{2.2.16}$$

$$T_{ac'bd'} = \frac{1}{2\pi}\chi_{abk}\bar{\chi}_{c'd'k}, \tag{2.2.17}$$

respectively. The continuity equation has the form

$$\partial_{ab'}J^{ab'}{}_{k} + J^{ab'}{}_{k}[(B_{cb'})_{a}{}^{c} + (B^{\dagger}_{f'a})^{f'}{}_{b'}] = 0. \tag{2.2.18}$$

After the tetrad components of the Yang-Mills field equations and variables are written down, one has to choose a coordinate system and an appropriate null tetrad. It is convenient to choose polar coordinates, in which the line element is given by

$$ds^2 = du^2 + 2\,du\,dr - r^2(d\theta^2 + \sin^2\theta\,d\phi^2), \qquad (2.2.19)$$

where $u = t - r$ is the retarded time. The surfaces $u = $ const are the light cones emanating from the origin $r = 0$.

At any point in space, one can choose the tetrad so that l^μ is the outward real null vector tangent to the light cone, n^μ is the inward real null vector pointing toward the origin, m^μ and $\bar{m}^\mu$ are complex null vectors tangent to the two-dimensional sphere defined by constants r and u. In the null coordinates

$$x^0 = u, \qquad x^1 = r, \qquad x^2 = \theta, \qquad x^3 = \phi,$$

these vectors are given by

$$l^\mu = \delta_1^\mu, \qquad (2.2.20a)$$

$$m^\mu = \frac{1}{\sqrt{2r}}\left(\delta_2^\mu + \frac{i}{\sin\theta}\delta_3^\mu\right), \qquad (2.2.20b)$$

$$n^\mu = \delta_0^\mu - \tfrac{1}{2}\delta_1^\mu. \qquad (2.2.20c)$$

It is easy to see that with this choice of the tetrad, the spin coefficients will take the following values [117]:

$$\varepsilon = \kappa = \pi = \sigma = \lambda = \gamma = \tau = \nu = 0, \qquad (2.2.21)$$

$$\rho = 2\mu = -\frac{1}{r}, \qquad \beta = -\alpha = \frac{1}{2\sqrt{2r}}\cot\theta. \qquad (2.2.22)$$

Using these values in the relations between the fields and the potentials, Eq. (2.2.11), one gets

$$\chi_{0k} = \frac{1}{\sqrt{2}r}\mathscr{D}b_{00'k} - \left(\frac{\partial}{\partial r} + \frac{1}{r}\right)b_{01'k} + g\varepsilon_{ijk}b_{01'i}b_{00'j}, \tag{2.2.23}$$

$$\chi_{1k} = \frac{1}{2}\left\{\left(\frac{\partial}{\partial u} - \frac{1}{2}\frac{\partial}{\partial r}\right)b_{00'k} + \frac{1}{\sqrt{2}r}[(\mathscr{D} + \cot\theta)b_{10'k}\right.$$

$$\left. - (\bar{\mathscr{D}} + \cot\theta)b_{01'k}] - \frac{\partial}{\partial r}b_{11'k} + g\varepsilon_{ijk}(b_{11'i}b_{00'j} - b_{10'i}b_{01'j})\right\}, \tag{2.2.24}$$

$$\chi_{2k} = \left(\frac{\partial}{\partial u} - \frac{1}{2}\frac{\partial}{\partial r} - \frac{1}{2r}\right)b_{10'k} - \frac{1}{\sqrt{2}r}\bar{\mathscr{D}}b_{11'k} + g\varepsilon_{ijk}b_{11'i}b_{10'j}. \tag{2.2.25}$$

In the last equations, use is made of the following notations:

$$\chi_{00k} = \chi_{0k}, \qquad \chi_{01k} = \chi_{10k} = \chi_{1k}, \qquad \chi_{11k} = \chi_{2k},$$

$$\tag{2.2.26}$$

$$\mathscr{D} = \frac{\partial}{\partial\theta} + \frac{i}{\sin\theta}\frac{\partial}{\partial\phi},$$

and $\bar{\mathscr{D}}$ is the complex conjugate of $\mathscr{D}$.

Substituting now the spin coefficients from Eqs. (2.2.21) and (2.2.22) in the Yang-Mills field equations (2.2.15), the latter will then take the following form:

$$\left(\frac{\partial}{\partial r} + \frac{2}{r}\right)\chi_{1k}$$

$$= \frac{1}{\sqrt{2}r}(\bar{\mathscr{D}} + \cot\theta)\chi_{0k} - g\varepsilon_{ijk}(\chi_{0i}b_{10'j} - \chi_{1i}b_{00'j}) + 2\pi j_{00'k}, \tag{2.2.27}$$

$$\left(\frac{\partial}{\partial u} - \frac{1}{2}\frac{\partial}{\partial r} - \frac{1}{2r}\right)\chi_{0k}$$

$$= \frac{1}{\sqrt{2}r}\mathscr{D}\chi_{1k} + g\varepsilon_{ijk}(\chi_{0i}b_{11'j} - \chi_{1i}b_{01'j}) + 2\pi j_{01'k}, \tag{2.2.28}$$

$$\left(\frac{\partial}{\partial r} + \frac{1}{r}\right)\chi_{2k}$$

$$= \frac{1}{\sqrt{2}r}\bar{\mathscr{D}}\chi_{1k} - g\varepsilon_{ijk}(\chi_{1i}b_{10'j} - \chi_{2i}b_{00'j}) + 2\pi j_{10'k},\qquad (2.2.29)$$

$$\left(\frac{\partial}{\partial u} - \frac{1}{2}\frac{\partial}{\partial r} - \frac{1}{r}\right)\chi_{1k}$$

$$= \frac{1}{\sqrt{2}r}(\mathscr{D} + \cot\theta)\chi_{2k} + g\varepsilon_{ijk}(\chi_{1i}b_{11'j} - \chi_{2i}b_{01'j}) + 2\pi j_{11'k}.\quad (2.2.30)$$

Equations (2.2.27)–(2.2.30) are generalizations of the Maxwell equations written by means of the null tetrad [117], the generalization is expressed by the appearance of the cross-product terms of χ and b [106, 122].

The equations (2.2.23)–(2.2.25), describing the relations between the fields χ_{abk} and the potentials $b_{ab'k}$, and the field equations (2.2.27)–(2.2.30), as it can be seen, point out the symmetry of the Yang-Mills field equations and thus it is easier to work with them to find the exact solutions. Using the above equations, new exact solutions were obtained [90, 91, 93].

In the next chapter we present different methods to classify the Yang-Mills field.

3
Classification of the Yang-Mills Fields

In this chapter we review three known classification methods for the Yang-Mills fields [94–102]. The problem of the classification of fields is deeply related to exact solutions describing the fields. Each exact solution belongs to a type of field according to the classification scheme. Thus the classification gives a deep insight into the physics of these fields, and contributes to their understanding. We first classify the electromagnetic field, and then go to the Yang-Mills fields. Some of the classification methods are Lorentz invariant, while the others are gauge invariant. The relation between the different methods will be given.

The first method presented is that of Carmeli [94–100], and is based on the different eigenspinors and eigenvalues that describe the Yang-Mills fields. The second is the Wang and Yang method [101], in which the classification is made by means of the rank of the matrix that describes the fields. The third is that of Castillejo, Kugler and Roskies [102], in which the fields are classified according to the little group that leaves them invariant. In the sequel, we will present each one of these methods separately.

3.1 *The eigenspinor-eigenvalue method*

3.1.1 *The electromagnetic field invariants*

We start with the classification of the electromagnetic field which is given by the anti-symmetric tensor $f_{\mu\nu}$. From this tensor one can form two invariants under transformations between inertial coordinate systems. These invariants are given by [13]

$$f_{\mu\nu}f^{\mu\nu} = 4K_1, \qquad \text{(scalar invariant)} \qquad (3.1.1)$$

$$f_{\mu\nu}*f^{\mu\nu} = 4K_2. \qquad \text{(pseudoscalar invariant)} \qquad (3.1.2)$$

Here $*f^{\mu\nu}$ is the dual to the tensor $f_{\mu\nu}$ defined by

$$*f^{\mu\nu} = \tfrac{1}{2}\varepsilon^{\mu\nu\rho\sigma}f_{\rho\sigma}, \qquad (3.1.3)$$

where $\varepsilon^{\mu\nu\rho\sigma}$ is the anti-symmetric Levi-Civita tensor and whose values are $+1$ or -1, depending upon whether $\mu\nu\rho\sigma$ is an even or odd permutation of 0123, and zero otherwise.

It can be shown that K_1 and K_2 are the only invariants that can be formed from the electromagnetic field. They are also given by the strengths of the electric field **E** and the magnetic field **H** in the following form:

$$K_1 = \tfrac{1}{2}(H^2 - E^2), \qquad (3.1.4)$$

$$K_2 = -\mathbf{E}\cdot\mathbf{H}. \qquad (3.1.5)$$

There is another way to present these invariants when the electromagnetic field is described by the spinor ϕ. The only invariant one can obtain from the symmetrical spinor ϕ_{AB} is

$$K = \phi_{AB}\phi^{AB}, \qquad (3.1.6)$$

where Latin capital indices take the values 0, 1. It is easy to show [122] that

$$K = K_1 + iK_2. \tag{3.1.7}$$

At this point, we will give an explanation to part of the notations that we use in this section. In Chapter 2, we used tetrad components and the appropriate indices are the dyad indices a and b' which take the values 0, 1 and $0'$, $1'$, respectively. In Chapter 3, we will use dyad components which are equivalent to tetrad components by appropriate choice of the dyad [14]. The capital Latin indices A and B' are spinorial and they take the values 0, 1 and $0'$, $1'$, respectively. Raising or lowering a spinor index is made by means of the anti-symmetric spinor ε which was defined by Eq. (2.2.8), as follows:

$$\xi^A = \varepsilon^{AB}\xi_B, \qquad \eta^{A'} = \varepsilon^{A'B'}\eta_{B'}, \tag{3.1.8}$$

$$\xi_A = \xi^B\varepsilon_{BA}, \qquad \eta_{A'} = \eta^{B'}\varepsilon_{B'A'}. \tag{3.1.9}$$

In this notation the spinor which is analogous to the tensor $f_{\mu\nu}$ is given by

$$f_{AB'CD'} = \sigma^\mu_{AB'}\sigma^\nu_{CD'}f_{\mu\nu}, \tag{3.1.10}$$

where $\sigma^\mu_{AB'}$ are Hermitian matrices which satisfy

$$\sigma^\mu_{AB'}\sigma^{\nu AB'} = g^{\mu\nu}, \tag{3.1.11}$$

$$\sigma^\mu_{AB'}\sigma^{AB'}_\nu = \delta^\mu_\nu. \tag{3.1.12}$$

Thus each tensorial index is replaced by two spinorial indices, one of which is unprimed and the other is primed.

3.1.2 *The eigenspinor-eigenvalue equation*

Now we consider the following eigenspinor-eigenvalue equation for the electromagnetic field:

$$\phi^A{}_B\alpha^B = \lambda\alpha^A, \tag{3.1.13}$$

where ϕ_{AB} is the electromagnetic spinor, α^A and λ are the eigenspinor and the eigenvalue, respectively. A simple calculation shows that the equation for the eigenvalues obtained from Eq. (3.1.13) is given by

$$\lambda^2 - \phi^A{}_A \lambda - \tfrac{1}{2} \phi^A{}_B \phi^B{}_A = 0, \tag{3.1.14}$$

or (since ϕ is symmetric),

$$\lambda^2 + \frac{K}{2} = 0. \tag{3.1.15}$$

One then has

$$\lambda_1 + \lambda_2 = \phi^A{}_A = 0, \tag{3.1.16}$$

$$\lambda_1^2 + \lambda_2^2 = (\phi^A{}_A)^2 + \phi^A{}_B \phi^B{}_A = -K, \tag{3.1.17}$$

where λ_1 and λ_2 are the eigenvalues given by

$$\lambda_{1,2} = \pm \left(-\frac{K}{2} \right)^{1/2}. \tag{3.1.18}$$

These eigenvalues may or may not be distinct according to the vanishing or nonvanishing of the invariant K.

3.1.3 *Classification of the electromagnetic field*

Now, the electromagnetic field can be classified according to the possible number of distinct eigenvalues and eigenspinors. The maximum number of eigenvalues is two. For each eigenvalue there corresponds one eigenspinor. For the case of two distinct eigenvalues there are two distinct eigenspinors. This is the general case of type I field. When the two eigenvalues are equal, then by Eq. (3.1.18), they are zero. In this case, if there are two linearly independent eigenspinors, then from Eq. (3.1.13), the spinor ϕ_{AB} is necessarily zero, and this is type O. If for the zero eigenvalue there corresponds only one eigenspinor, then one has the case of a null electromagnetic field N. The three types are shown in Table 3.1.1.

The classification may also be described in a diagram (see Figure 3.1.1).

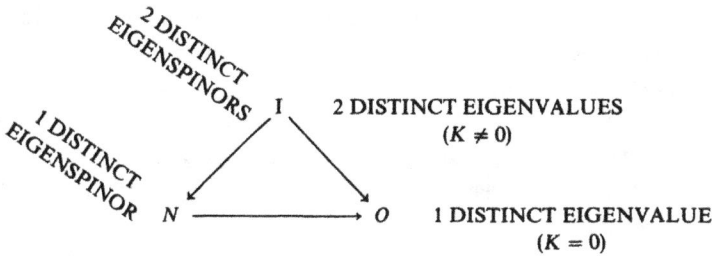

Fig. 3.1.1 The classification of the electromagnetic field in terms of the invariant K and the eigenspinors and the eigenvalues.

Table 3.1.1 The classification of the electromagnetic field according to distinct eigenspinors and eigenvalues.

Type of the electromagnetic field	Type I	Type O	Type N
Distinct eigenspinors	2	2	1
Distinct eigenvalues	2	1	1

Another way to see the above classification is by means of the decomposition of the spinor ϕ_{AB}. Since ϕ_{AB} is symmetric, one can factorize it as follows:

$$\phi_{AB} = \alpha_{(A}\beta_{B)}. \tag{3.1.19}$$

Here α_A and β_A are arbitrary spinors with one index. Consequently, the invariant K is given by

$$K = -\tfrac{1}{2}(\alpha_A\beta^A)^2. \tag{3.1.20}$$

Now, the classification of the electromagnetic field can be made in terms of this decomposition. If the two spinors α_A and β_A are distinct from each other, then $K \neq 0$, and this is the general case I. If the two spinors α_A and β_A are equal or proportional to each other, then $K = 0$ and the electromagnetic field is null or zero, respectively.

Similar methods were used to classify the Weyl tensor describing the gravitational field [122]. We will not review this topic, and we go to the classification of the SU(2) gauge fields, where the methods which will be used are similar to those used to classify the electromagnetic and the gravitational fields. In this case the structure of the fields depends on the SL(2, C) space-time group and the SU(2) internal space group, as compared to gravitation whose both space-time and internal groups are SL(2, C). This fact makes the classification problem of gauge fields of greater interest. Of course, one may classify gauge fields that are associated with larger groups than SU(2), but the techniques will be more complicated [103].

3.1.4 *The gauge-fields invariants*

To construct the invariants of the SU(2) gauge fields, one defines the following spinor from the Yang-Mills tensor $f_{\mu\nu k}$ (see Chapter 2)

$$f_{AB'CD'k} = \sigma^{\mu}_{AB'}\sigma^{\nu}_{CD'}f_{\mu\nu k}. \tag{3.1.21}$$

Here $f_{AB'CD'k}$ is the spinor that describes the Yang-Mills field. This spinor can be decomposed as follows:

$$f_{AB'CD'k} = \chi_{ACk}\varepsilon_{B'D'} + \varepsilon_{AC}\bar{\chi}_{B'D'k}, \tag{3.1.22}$$

where the symmetric spinor χ_{ABk} is given by

$$\chi_{ABk} = \tfrac{1}{2}\varepsilon^{C'D'}f_{AC'BD'k}. \tag{3.1.23}$$

Now, one defines the following spinor

$$\xi_{ABCD} = \chi_{ABk}\chi_{CDk}. \tag{3.1.24}$$

It can be shown that this spinor satisfies the following symmetry properties:

$$\xi_{ABCD} = \xi_{BACD} = \xi_{ABDC} = \xi_{CDAB}. \tag{3.1.25}$$

The spinor ξ_{ABCD} can be decomposed into a totally symmetric spinor and a scalar P such that

$$\xi_{ABCD} = \eta_{ABCD} + \frac{P}{6}(\varepsilon_{AC}\varepsilon_{BD} + \varepsilon_{AD}\varepsilon_{BC}), \tag{3.1.26}$$

where

$$p = \xi_{AB}{}^{AB} = \tfrac{1}{4} f_{\mu\nu k}(f^{\mu\nu}{}_k + i \, {}^*f^{\mu\nu}{}_k).$$ (3.1.27)

It can be shown that

$$\xi_{AC}{}^{C}{}_{B} = \frac{P}{2}\varepsilon_{AB}.$$ (3.1.28)

The symmetric spinor η_{ABCD} is given by

$$\eta_{ABCD} = \tfrac{1}{3}(\xi_{ABCD} + \xi_{ACBD} + \xi_{ADBC}).$$ (3.1.29)

The number of the real invariants of the SU(2) gauge fields, which one can construct, is nine. We will give the invariants that are used in the classification of the fields. More information about these invariants can be found in Ref. 95.

In addition to the invariant P which was defined by Eq. (3.1.27), one defines the following two invariants:

$$G = \eta_{ABCD}\eta^{ABCD},$$ (3.1.30)

$$H = \eta_{AB}{}^{CD}\eta_{CD}{}^{EF}\eta_{EF}{}^{AB}.$$ (3.1.31)

3.1.5 *The eigenspinor-eigenvalue equation*

The eigenspinor-eigenvalue equation is written in terms of the spinor ξ_{ABCD}, which is invariant under the gauge transformations, and is given by [95]

$$\xi^{AB}{}_{CD}\phi^{CD} = \lambda\phi^{AB},$$ (3.1.32)

where ϕ^{AB} is a symmetric eigenspinor. Using Eq. (3.1.26), in which the spinor ξ_{ABCD} is given by the total symmetric spinor η_{ABCD} and the invariant P, Eq. (3.1.32) will be

$$\eta^{AB}{}_{CD}\phi^{CD} = \lambda'\phi^{AB},$$ (3.1.33)

where the new eigenvalue λ' is related to λ by

$$\lambda' = \lambda - P/3.\tag{3.1.34}$$

According to Eq. (3.1.33), the classification of the spinor ζ_{ABCD} reduces to the classification of the completely symmetric spinor η_{ABCD}. The eigenvalue equation obtained from Eq. (3.1.33) will be

$$f(\lambda') \equiv \lambda'^3 - \tfrac{1}{2}G\lambda' - \tfrac{1}{3}H = 0,\tag{3.1.35}$$

where G and H are the invariants defined by Eqs. (3.1.30) and (3.1.31), respectively. Then one obtains

$$\lambda'_1 + \lambda'_2 + \lambda'_3 = \eta_{AB}{}^{AB} = 0,\tag{3.1.36}$$

$$\lambda'^2_1 + \lambda'^2_2 + \lambda'^2_3 = \eta_{ABCD}\eta^{ABCD} = G,\tag{3.1.37}$$

$$\lambda'^3_1 + \lambda'^3_2 + \lambda'^3_3 = \eta_{AB}{}^{CD}\eta_{CD}{}^{EF}\eta_{EF}{}^{AB} = H,\tag{3.1.38}$$

where λ'_1, λ'_2, and λ'_3 are the eigenvalues which may or may not be distinct.

The spinor η_{ABCD}, and consequently, the spinor ζ_{ABCD}, will be classified with respect to the number of the distinct eigenvalues and eigenspinors. The maximum number of eigenvalues is three. For each eigenvalue, there is at least one eigenspinor. Hence when there are three distinct eigenvalues, there are three eigenspinors. This is the general type I field. Since there are two cases for which $P \neq 0$ and $P = 0$, one obtains the fields of type I_p and I_o, respectively. The spinor η_{ABCD} will then have the general form

$$\eta_{ABCD} = \alpha_{(A}\beta_B\gamma_C\delta_{D)},\tag{3.1.39}$$

where α_A, β_B, γ_C and δ_D are four arbitrary one-index spinors, and the parentheses indicate symmetrization, thus there are 24 terms in Eq. (3.1.39). Analysis of Eqs. (3.1.36)–(3.1.38) shows that in this case, one has $G^3 \neq GH^2$.

When two of the eigenvalues, let us say λ'_1 and λ'_2, are equal, one has two or three distinct eigenspinors. The fields will be of types II and D, respectively. Also here, for each type there are two cases, $P \neq 0$ and $P = 0$. In this case, the spinor η_{ABCD} will take the form such that two of the one-index spinors are identical,

$$\eta_{ABCD} = \alpha_{(A}\alpha_B\gamma_C\delta_{D)},\tag{3.1.40}$$

for the types II_p and II_0. For types D_p and D_0, the four spinors are identical in pairs,

$$\eta_{ABCD} = \alpha_{(A}\alpha_B\delta_C\delta_{D)}. \tag{3.1.41}$$

From Eqs. (3.1.36)–(3.1.38), one sees that $G^3 = GH^2 \neq 0$ for these four cases.

Finally, if $\lambda_1' = \lambda_2' = \lambda_3'$, then one may have one, two or three eigenspinors. The fields obtained are of types III, IV or 0, respectively. The spinor η_{ABCD} will then have three equal spinors out of the four,

$$\eta_{ABCD} = \alpha_{(A}\alpha_B\alpha_C\delta_{D)}, \tag{3.1.42}$$

for the fields of types III_p and III_0. The four spinors are equal,

$$\eta_{ABCD} = \alpha_A\alpha_B\alpha_C\alpha_D, \tag{3.1.43}$$

for the fields of types IV_p, IV_0. And finally, $\eta_{ABCD} = 0$ for the fields of types 0_p and 0_0. From Eqs. (3.1.36)–(3.1.38), one obtains $G = H = 0$ for the last six types. A summary is given in Fig. 3.1.2.

One can use the spinor ξ_{ABCD} to define the following invariants [98]

$$S = \xi_{ABCD}\xi^{ABCD} = G + \tfrac{1}{3}P^2, \tag{3.1.44}$$

$$F = \xi_{AB}{}^{CD}\xi_{CD}{}^{EF}\xi_{EF}{}^{AB} = H + PG + \tfrac{1}{9}P^3. \tag{3.1.45}$$

Then, using these invariants and Eq. (3.1.34) in the eigenvalues Eq. (3.1.35), one obtains

$$\lambda^3 + m_1\lambda^2 + m_2\lambda + m_3 = 0, \tag{3.1.46}$$

where

$$m_1 = -P, \tag{3.1.47}$$

$$m_2 = \tfrac{1}{2}(P^2 - S), \tag{3.1.48}$$

$$m_3 = -\tfrac{1}{3}(F - \tfrac{3}{2}PS + \tfrac{1}{2}P^3). \tag{3.1.49}$$

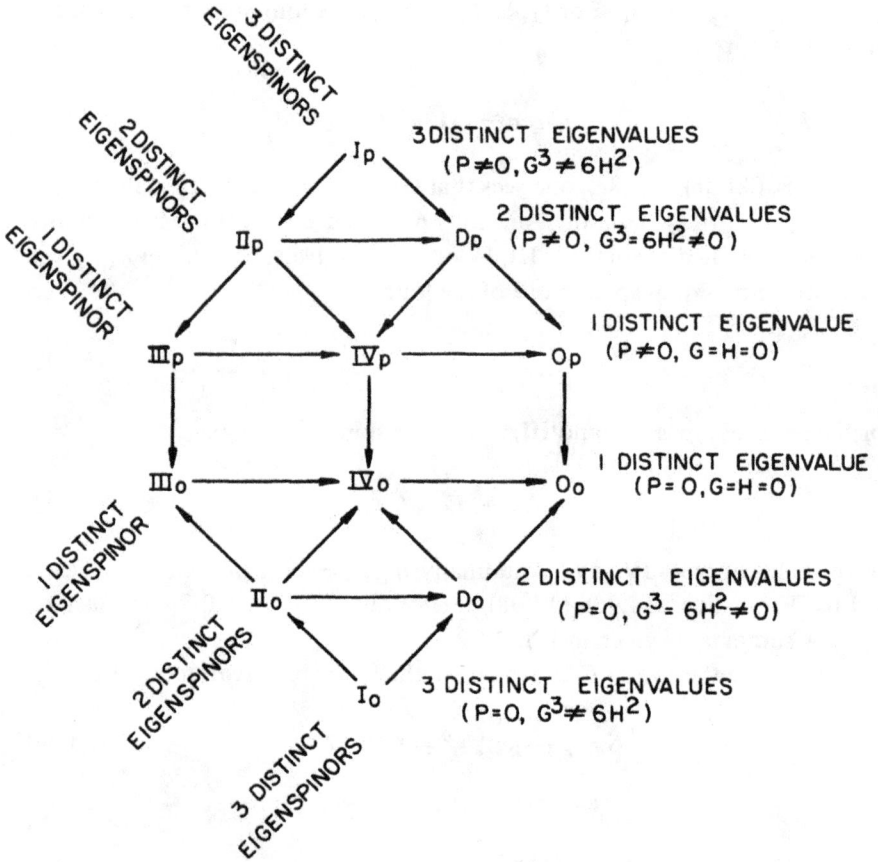

Fig. 3.1.2 Diagram of classification of SU(2) gauge fields. The completely symmetrical spinor η_{ABCD} is decomposed for each one of the twelve types of fields in the diagram. For types I_p and I_0 one has $\eta_{ABCD} = \alpha_{(A}\beta_B\gamma_C\delta_{D)}$, where the two-component spinors α, β, γ and δ are distinct from each other; for types II_p and II_0 two from the four spinors coincide; for types D_p and D_0 the four spinors coincide in pairs; for types III_p and III_0 three spinors coincide; for types IV_p and IV_0 the four spinors coincide; for types 0_p and 0_0 the spinor η vanishes. (Notice that for the fields I_p, II_p, etc., the invariant $P \neq 0$, whereas for I_0, II_0, etc., $P = 0$.)

The invariants P, S and F can also be defined by

$$P = \mathrm{Tr}\,\xi = \mathrm{Tr}\,\Delta, \qquad (3.1.50)$$

$$S = \mathrm{Tr}\,\xi^2 = \mathrm{Tr}\,\Delta^2, \qquad (3.1.51)$$

$$F = \mathrm{Tr}\,\xi^3 = \mathrm{Tr}\,\Delta^3, \qquad (3.1.52)$$

where ξ is a symmetrical matrix, which is invariant under the gauge group, obtained from the spinor ξ_{ABCD} when expanded in an appropriate basis in the spin space, and Δ is a symmetrical matrix whose elements are given by

$$\Delta_{ij} = \chi_{iAB}\chi_j^{AB}. \qquad (3.1.53)$$

The matrix Δ is Lorentz invariant. Wang and Yang used the rank of Δ to classify the gauge fields. It can be shown that

$$m_3 = -\det\xi = -\det\Delta = \frac{1}{54}(9PG - 18H - 2P^3). \qquad (3.1.54)$$

The two matrices Δ and ξ have a simple presentation. If $A_i^k = E_i^k + iH_i^k$, where E_i^k and H_i^k are the electric and the magnetic parts of the Yang-Mills fields, then $A_i^k = e_i^{AB}\chi_{AB}^k$ where e_i^{AB} is an appropriate basis in the spinor space, and $\Delta = A^tA$ and $\xi = AA^t$. Their transformation rules are then given by $A' = AG$ and $A' = LA$, A^t is the transposed matrix, thus

$$\Delta' = G^t\Delta G, \qquad \xi' = L\xi L^t, \qquad (3.1.55)$$

where G is a three-dimensional orthogonal real matrix and L is a three-dimensional complex orthogonal matrix, both with determinants unity.

Finally, one notices that the eigenvalues λ_1, λ_2 and λ_3 satisfy

$$\sum \lambda_i = \xi_{AB}^{\ \ AB} = P, \qquad (3.1.56)$$

$$\sum \lambda_i^2 = \xi_{ABCD}\xi^{ABCD} = S, \qquad (3.1.57)$$

$$\sum \lambda_i^3 = \xi_{AB}^{\ \ CD}\xi_{CD}^{\ \ EF}\xi_{EF}^{\ \ AB} = F. \qquad (3.1.58)$$

The canonical form for the matrix corresponding to the spinor ξ_{ABCD} in a Lorentz frame can be given, along with the components of the spinor η_{ABCD} [122].

In the following section we will give the classification of the SU(2) gauge fields according to the method of the rank of the matrix.

3.2 *The matrix method*

From the gauge field strengths, one can construct invariants of the gauge group which are not invariant under the Lorentz group, or the opposite, invariants of the Lorentz group which are not invariant under the gauge group. In the last section, the gauge fields were classified by means of SU(2) gauge invariants and eigenspinor-eigenvalue equations which are not Lorentz invariant, a method that was developed by Carmeli. In this section, the classification will be done by means of Lorentz invariants which are not SU(2) gauge invariant. This method was developed by Wang and Yang [101], in which the rank of the matrix was used as a tool for the classification of the fields. One starts with the following questions: Given the Lorentz invariants of the fields, are they realizable? If they are, then how many inequivalent realizations exist? (Two realizations are called inequivalent if they are not obtained one from the other by Lorentz and gauge transformations. A realization is called unique if there exists only one inequivalent realization.) Can one choose some standard forms of realization?

Gauge fields are represented by 3×3 complex matrices, the columns of which can be viewed as electromagnetic fields. It therefore, proves useful to discuss first the classification scheme of electromagnetic fields (see Wang and Yang, Ref. 101).

3.2.1 *Electromagnetic fields*

If E and H are real column vectors

$$E = \begin{pmatrix} E_x \\ E_y \\ E_z \end{pmatrix}, \qquad H = \begin{pmatrix} H_x \\ H_y \\ H_z \end{pmatrix} \qquad (3.2.1)$$

representing the electric and magnetic field strengths, then the complex column vector

$$A = E + iH = \begin{pmatrix} A_x \\ A_y \\ A_z \end{pmatrix} \qquad (3.2.2)$$

represents the combined electromagnetic field. Under a Lorentz transformation A is transformed by a 3×3 complex orthogonal matrix L with unit determinant:

$$A \to LA. \qquad (3.2.3)$$

One constructs the Lorentz-invariant

$$\Delta = A^t A = K + iJ, \qquad (3.2.4)$$

which is a 1×1 matrix, i.e., a complex number, and serves as a tool for classifying electromagnetic fields. Given an arbitrary value of Δ, it is always realized by an electromagnetic field, and the number of the Lorentz-inequivalent realizations is determined according to the two distinct possibilities:

Case 1. $\Delta \neq 0$ *(Type I of Sec. 3.1)*

Every realization is equivalent to the standard realization

$$A = \begin{pmatrix} a \\ 0 \\ 0 \end{pmatrix} \qquad a = \text{positive complex} = \sqrt{\Delta} \qquad (3.2.5)$$

(a complex number is called positive if either its real part is positive, or it is purely imaginary and its imaginary part is positive.)

Case 2. $\Delta = 0$ *(Types N and O of Sec. 3.1)*

Every realization is equivalent to one of the two inequivalent standard realizations

$$A = \begin{pmatrix} 1 \\ i \\ 0 \end{pmatrix} \qquad \text{(Null field, type N of Sec. 3.1)} \qquad (3.2.6)$$

and

$$A = \begin{pmatrix} 0 \\ 0 \\ 0 \end{pmatrix} \quad \text{(Zero field, type } O \text{ of Sec. 3.1)} \qquad (3.2.7)$$

3.2.2 SU(2) *gauge fields*

The Wang-Yang classification of SU(2) gauge fields [101] is modeled after the analysis of electromagnetic fields in the foregoing discussion. The gauge field strengths E^k and H^k, with the isospin index k assuming the values 1, 2, 3, are written in matrices:

$$E = \begin{pmatrix} E^1_x & E^2_x & E^3_x \\ E^1_y & E^2_y & E^3_y \\ E^1_z & E^2_z & E^3_z \end{pmatrix} \qquad H = \begin{pmatrix} H^1_x & H^2_x & H^3_x \\ H^1_y & H^2_y & H^3_y \\ H^1_z & H^2_z & H^3_z \end{pmatrix}, \qquad (3.2.8)$$

and they are combined into the complex 3×3 matrix

$$A = E + iH.$$

The gauge field A is subject to two kinds of transformations: a Lorentz transformation

$$A \rightarrow LA,$$

with L being a 3×3 complex orthogonal matrix with unit determinant, and a local gauge transformation

$$A \rightarrow AG,$$

with G being a 3×3 real orthogonal matrix with unit determinant. The Lorentz invariant in the present case is the 3×3 complex symmetric matrix

$$\Delta = A^t A = K + iJ. \qquad (3.2.9)$$

Under a gauge transformation G, this quadratic Lorentz invariant transforms as

$$\Delta \rightarrow G^t \Delta G.$$

The rank of the matrix Δ serves as the tool for classifying SU(2) gauge fields according to the following scheme [101]:

Case 1. *rank* $\Delta = 3$

There exists a gauge ("proper gauge") in which the elements Δ^{kj} of the matrix Δ satisfy

$$\Delta^{11} \neq 0, \quad \det \begin{pmatrix} \Delta^{11} & \Delta^{12} \\ \Delta^{21} & \Delta^{22} \end{pmatrix} \neq 0. \tag{3.2.10}$$

In a proper gauge, every realization is Lorentz-equivalent to one of the two gauge-Lorentz-inequivalent standard realizations

$$A = \begin{pmatrix} a & f & e \\ 0 & b & d \\ 0 & 0 & c \end{pmatrix} \quad \text{and} \quad A = \begin{pmatrix} a & f & e \\ 0 & b & d \\ 0 & 0 & -c \end{pmatrix}, \tag{3.2.11}$$

where $a, \ldots, f$ are complex numbers, with a, b and c being positive complex, all these six complex numbers are determined uniquely by the elements of Δ in the proper gauge.

Case 2. *rank* $\Delta = 2$

There exists a gauge ("proper gauge") in which the condition (3.2.10) is satisfied. In a proper gauge, every realization is Lorentz-equivalent to the standard realization

$$A = \begin{pmatrix} a & f & e \\ 0 & b & d \\ 0 & 0 & 0 \end{pmatrix}, \tag{3.2.12}$$

where $a, \ldots, f$ are complex numbers, with a and b being positive complex, all these five complex numbers are determined uniquely by the elements of Δ in the proper gauge.

Case 3. rank $\Delta = 1$

There exists a gauge (a "proper gauge") in which the matrix Δ assumes the form

$$\Delta = \begin{pmatrix} a \\ f \\ 0 \end{pmatrix} (a \quad f \quad 0), \tag{3.2.13}$$

where a and f are complex numbers, a being positive complex. In a proper gauge, every realization is Lorentz-gauge-equivalent to a realization of the form

$$A = \begin{pmatrix} a & f & 0 \\ 0 & \lambda & \mu \\ 0 & \lambda i & \mu i \end{pmatrix}, \tag{3.2.14}$$

where λ and μ are complex numbers. Notice that the gauge transformation employed in the passage to (3.2.14) does not take us out of a proper gauge. The classification of the latter falls into two subcases:

Subcase 3a. rank $\Delta = 1$, $f/a \neq real$

In a proper gauge in this subcase, every realization is Lorentz-gauge-equivalent to one of the following set of standard realizations:

$$A = \begin{pmatrix} a & f & 0 \\ 0 & \lambda & 1 \\ 0 & \lambda i & i \end{pmatrix}, \quad A = \begin{pmatrix} a & f & 0 \\ 0 & 1 & 0 \\ 0 & i & 0 \end{pmatrix}, \quad A = \begin{pmatrix} a & f & 0 \\ 0 & 0 & 0 \\ 0 & 0 & 0 \end{pmatrix}, \tag{3.2.15}$$

with λ being an arbitrary complex number. Realizations of different types in (3.2.15), as well as those corresponding to the first type for distinct values of λ, are Lorentz-gauge-inequivalent.

Subcase 3b. rank $\Delta = 1$, $f/a = real$

In a proper gauge in this subcase, every realization is Lorentz-gauge-equivalent to one of the following set of standard realizations:

$$A = \begin{pmatrix} a & 0 & 0 \\ 0 & i & \lambda \\ 0 & -1 & \lambda i \end{pmatrix}, \qquad A = \begin{pmatrix} a & 0 & 0 \\ 0 & 0 & 0 \\ 0 & 0 & 0 \end{pmatrix}, \qquad (3.2.16)$$

where λ is a real number, $-1 \leq \lambda \leq 1$. Realizations of different types in (3.2.16), as well as those corresponding to the first type for distinct values of λ, are Lorentz-gauge-inequivalent.

Case 4. $\Delta = 0$

Every realization is Lorentz-gauge-equivalent to one of the following set of standard realizations:

$$A = \begin{pmatrix} 0 & 0 & 0 \\ 0 & i & \lambda \\ 0 & -1 & \lambda i \end{pmatrix}, \qquad A = 0, \qquad (3.2.17)$$

where λ is a real number, $0 \leq \lambda \leq 1$. Realizations of different types in (3.2.17), as well as those corresponding to the first type for distinct values of λ, are Lorentz-gauge-inequivalent.

A comparison of the two classification schemes—the Lorentz invariant method and the gauge invariant method—will be presented in the next section.

3.3 *Lorentz invariance versus gauge invariance—comparison*

In the last two sections, we presented two methods of classification of SU(2) gauge fields—one method is gauge invariant and the second is Lorentz invariant. Now we present a comparison between these two methods [98].

To compare the two methods, one starts with the case for which the ranks of the matrices Δ and ξ are equal to three, namely their determinants are different from zero. From Eq. (3.1.54), one can see that the polynomial

$$m_3(P) = -\tfrac{1}{27}P^3 + \tfrac{1}{6}GP - \tfrac{1}{3}H, \tag{3.3.1}$$

which is of third order in the invariant P, does not vanish. By changing variables from P to Z such that $Z = -P/3$, one obtains the polynomial

$$f(Z) = Z^3 - \tfrac{1}{2}GZ - \tfrac{1}{3}H. \tag{3.3.2}$$

This is exactly the polynomial one obtains from the eigenspinor equation (3.1.32), as it should be.

From the above it is clear that there exists a gauge in which Δ is realized by a matrix whose elements are constructed out of the invariants P, G and H. A possible presentation of such a matrix is given by

$$\Delta = \begin{pmatrix} \tfrac{1}{3}P & \alpha_3 & \alpha_2 \\ \alpha_3 & \tfrac{1}{3}P & \alpha_1 \\ \alpha_2 & \alpha_1 & \tfrac{1}{3}P \end{pmatrix}, \tag{3.3.3}$$

where $\sum \alpha_i^2 = G/2$ and $\alpha_1 \alpha_2 \alpha_3 = H/6$. One may easily check that $\mathrm{Tr}\,\Delta = P$, $\mathrm{Tr}\,\Delta^2 = P^2/3 + G$, $\mathrm{Tr}\,\Delta^3 = P^3/9 + PG + H$, and $\det \Delta$ satisfies Eq. (3.1.54).

Coming back to the polynomial (3.3.1) and the condition $\det \Delta = (2P^3 - 9GP + 18H)/54 \neq 0$. This is case 1 in the matrix method of classification. It yields eight subcases [98]. They correspond to nine types of fields in the eigenspinor-eigenvalue classification scheme. These are I_p, II_p, D_p, III_p, IV_p, 0_p, I_0, II_0 and D_0. When $\det \Delta = 0$ and rank $\Delta = 2$, which is case 2 in the matrix classification scheme, one has six subcases [98]. They correspond to five types of fields in the eigenspinor-eigenvalue classification scheme. These are I_0, IV_0, I_p, II_p and D_p. Summaries of the correspondence between the two methods are given in Tables 3.3.1 and 3.3.2 [98], including the cases for which the ranks of $\Delta = 1$ and 0.

Table 3.3.1 Corresponding cases using the rank of the matrix Δ (cases 1, 2, 3, 4 denote the ranks of $\Delta = 3, 2, 1, 0$) versus field types using the eigenspinor equation.

Matrix method case	Corresponding types in eigenspinor-eigenvalue method
1	I_p, II_p, D_p, III_p, IV_p, 0_p, I_0, II_0, D_0
2	I_p, II_p, D_p, I_0, IV_0
3	II_p, D_p, III_0, IV_0, 0_0
4	IV_0, 0_0

Table 3.3.2 Types of fields using the eigenspinor equation and their corresponding cases using the rank of the matrix Δ (cases 1, 2, 3, 4 denote ranks of Δ equal to 3, 2, 1, 0).

Eigenspinor-eigenvalue method type	Corresponding case in the matrix method
I_p	1, 2
II_p, D_p	1, 2, 3
III_p, IV_p, O_p	1
I_0	1, 2
II_0, D_0	1
III_0	3
IV_0	2, 3, 4
O_0	3, 4

3.4 *The little-group method*

In the first two sections of this chapter, we reviewed two classification methods for the SU(2) gauge fields. One of them was based on the gauge-invariant eigenspinor equation, the other, in which the rank of the matrix was used as a tool to classify the fields, was Lorentz invariant. In this section we will review another classification method for the gauge fields that is different from these two schemes. In this method, which was developed by Castillejo, Kugler and Roskies, one uses the subgroups of the Lorentz and the gauge groups, which leave the gauge fields invariant, as a tool to classify the fields [102].

Practically, one has to answer the following question: Which subgroup of the Lorentz and the gauge group leaves the gauge field invariant? And then one looks for the obtained little groups (subgroups) which are inequivalent. The construction of the little groups with field invariants give the classification of the fields and the standard form of the Yang-Mills fields which is Lorentz and gauge invariant. To understand this method, we first classify the electromagnetic field.

3.4.1 *The electromagnetic field*

In the first two sections of this chapter, the electromagnetic field was classified by two methods, the first was gauge invariant and the second was Lorentz invariant. In the first method the field was classified into three types, a general type where the field invariants are different from zero, and two other types where the field invariants vanish. In the second method, the field was classified into two cases. In the first case (which corresponds to the general type of the first method), the electric and the magnetic fields are parallel and have the form (3.2.5). The second case (which corresponds to the last two types of the first method) has two subcases in the first of which the electric and the magnetic fields are perpendicular to each other and have the form (3.2.6), and in the second subcase the fields vanish.

According to the method of Castillejo, Kugler and Roskies, the electromagnetic field is classified into three classes which differ from each other by the little group leaving the field invariant. These classes, denoted by I, II and III, correspond to types I, N and 0 of the Carmeli classification scheme. The little groups which leave the field invariant in each of these classes are given by [102]:

$$\text{Class I:} \quad \{G\} = \{J_1, K_1\}, g = U(1) \times E(1), \tag{3.4.1}$$

$$\text{Class II:} \quad \{G\} = \{K_1 + J_2, K_2 - J_1\}, g = E(1) \times E(1), \tag{3.4.2}$$

$$\text{Class III:} \quad \{G\} = \{K_1, K_2, K_3, J_1, J_2, J_3\}, g = \text{SL}(2, C), \tag{3.4.3}$$

where $U(1)$ is the one-dimensional unitary group and $E(1)$ is the one-dimensional Euclidean group, J_i and K_i, $i = 1, 2, 3$, are the generators of the rotations in three-dimensional space and the boosts (K_1 and K_2 in this section are not to be confused with the invariants K_1 and K_2 defined in Sec. 3.1).

One sees that the invariants of classes II and III are identical, but their little groups are different. It is obvious that the little group of class III is the whole Lorentz group. All equivalent fields have the same invariants and little groups. It is worthwhile mentioning that not every subgroup of $\text{SL}(2, C)$ is a little group of the electromagnetic field.

3.4.2 *Classification of the Yang-Mills fields*

Now we give the classification of the Yang-Mills fields according to the little group method. The fields will be described by the two matrices (3.2.8). (Our matrices are the transposed of those given in Ref. 102.)

The summation on Lorentz indices will give the symmetrical matrices K_{ij} and J_{ij} that were defined by Eq. (3.2.9). As was pointed out, from these matrices, one can construct the field invariants, and according to the rank of the matrix

$K + iJ$, Wang and Yang classified the Yang-Mills fields. In the little group method, the classification is done as follows: One takes a general system of inequivalent $SL(2, C) \times SU(2)$ generators, namely, generators which do not go to each other by Lorentz and gauge transformations. Then one constructs the general field which remains invariant under these generators. This is done by solving a linear system of equations. Finally, the fields are transformed to a standard form.

The inequivalent Lorentz generators are given by [102]

$$G_1 = J_1 \cos \theta + K_1 \sin \theta, \tag{3.4.4}$$

$$L_1 = K_1 + J_2, \tag{3.4.5}$$

where J_i and K_i are the generators of the rotations and the boosts, respectively. All the generators I_n of the isospin are equivalent. Therefore, the inequivalent $SL(2, C) \times SU(2)$ group generators are

$$I_1, G_1, L_1 \tag{3.4.6}$$

and

$$G_1 \cos \alpha + I_1 \sin \alpha, \qquad L_1 \cos \beta + I_1 \sin \beta. \tag{3.4.7}$$

The first three are Lorentz or gauge generators and the last two are combinations of Lorentz and gauge generators. Taking α or β equal to 0 or $\pi/2$ in Eq. (3.4.7), one obtains the generators of Eq. (3.4.6). Thus, the separation between the two cases is done for pedagogical reasons.

Given the generators, one then constructs the general field which is invariant under these generators. The field is subsequently transformed to the standard form. According to this method, one obtains eight classes of fields E_i and H_i which are invariant under the $SL(2, C) \times SU(2)$ group generators. The results are given in Table 3.4.1. More details on this classification can be found in Ref. 102. (The matrices which describe E and H in Table 3.4.1 are the transposed of those given in Ref. 102.)

Comparison between the little group and the spinorial methods is summarized in Tables 3.4.2 and 3.4.3.

In the next chapter we give some solutions of the Yang-Mills field equations.

Table 3.4.1 Classes of Yang-Mills fields using the little-group method. (The matrices E and H are the transposed of those of Ref. 102.)

Class	Generators	Canonical form of E	Canonical form of H	Matrix K	Matrix J	Rank of $K+iJ$
A	J_1, K_1	$\begin{pmatrix} a & 0 & 0 \\ 0 & a & 0 \\ 0 & 0 & 0 \end{pmatrix}$	$\begin{pmatrix} b & c & 0 \\ 0 & b & 0 \\ 0 & 0 & 0 \end{pmatrix}$	$\begin{pmatrix} a^2-b^2 & -bc & 0 \\ -bc & -c^2 & 0 \\ 0 & 0 & 0 \end{pmatrix}$	$\begin{pmatrix} 2ab & ac & 0 \\ ac & 0 & 0 \\ 0 & 0 & 0 \end{pmatrix}$	1
B	I_1, J_1, K_1	$\begin{pmatrix} a & 0 & 0 \\ 0 & a & 0 \\ 0 & 0 & 0 \end{pmatrix}$	$\begin{pmatrix} b & 0 & 0 \\ 0 & b & 0 \\ 0 & 0 & 0 \end{pmatrix}$	$\begin{pmatrix} a^2-b^2 & 0 & 0 \\ 0 & 0 & 0 \\ 0 & 0 & 0 \end{pmatrix}$	$\begin{pmatrix} 2ab & 0 & 0 \\ 0 & 0 & 0 \\ 0 & 0 & 0 \end{pmatrix}$	1
C_1	$I_1 + J_1$	$\begin{pmatrix} a & 0 & 0 \\ 0 & b & 0 \\ 0 & 0 & b \end{pmatrix}$	$\begin{pmatrix} c & 0 & 0 \\ 0 & d & 0 \\ 0 & 0 & d \end{pmatrix}$	$\begin{pmatrix} a^2-c^2 & 0 & 0 \\ 0 & b^2-d^2 & 0 \\ 0 & 0 & b^2-d^2 \end{pmatrix}$	$\begin{pmatrix} 2ac & 0 & 0 \\ 0 & 2bd & 0 \\ 0 & 0 & 2bd \end{pmatrix}$	3, 2
C_2	$I_1 + J_1$	$\begin{pmatrix} a & 0 & 0 \\ 0 & 1 & 0 \\ 0 & 0 & 1 \end{pmatrix}$	$\begin{pmatrix} b & 0 & 0 \\ 0 & 0 & -1 \\ 0 & 1 & 0 \end{pmatrix}$	$\begin{pmatrix} a^2-b^2 & 0 & 0 \\ 0 & 0 & 0 \\ 0 & 0 & 0 \end{pmatrix}$	$\begin{pmatrix} 2bd & 0 & 0 \\ 0 & 0 & 0 \\ 0 & 0 & 0 \end{pmatrix}$	1
D	$K_1 + J_2,$ $K_2 - J_1$	$\begin{pmatrix} a & 0 & 0 \\ 0 & 1 & 0 \\ 0 & 0 & 0 \end{pmatrix}$	$\begin{pmatrix} 0 & 1 & 0 \\ -a & 0 & 0 \\ 0 & 0 & 0 \end{pmatrix}$	$\begin{pmatrix} 0 & 0 & 0 \\ 0 & 0 & 0 \\ 0 & 0 & 0 \end{pmatrix}$	$\begin{pmatrix} 0 & 0 & 0 \\ 0 & 0 & 0 \\ 0 & 0 & 0 \end{pmatrix}$	0
E	$I_1, K_1 + J_2,$ $K_2 - J_1$	$\begin{pmatrix} 1 & 0 & 0 \\ 0 & 0 & 0 \\ 0 & 0 & 0 \end{pmatrix}$	$\begin{pmatrix} 0 & 0 & 0 \\ 0 & 1 & 0 \\ 0 & 0 & 1 \end{pmatrix}$	$\begin{pmatrix} 0 & 0 & 0 \\ 0 & 0 & 0 \\ 0 & 0 & 0 \end{pmatrix}$	$\begin{pmatrix} 0 & 0 & 0 \\ 0 & 0 & 0 \\ 0 & 0 & 0 \end{pmatrix}$	0
F	$K_1 + J_2,$ $K_2 - J_1$	$\begin{pmatrix} 1 & 0 & 0 \\ 0 & 1 & 0 \\ 0 & 0 & 0 \end{pmatrix}$	$\begin{pmatrix} -1 & 0 & 0 \\ 0 & 1 & 0 \\ 0 & 0 & 0 \end{pmatrix}$	$\begin{pmatrix} 0 & 0 & 0 \\ 0 & 0 & 0 \\ 0 & 0 & 0 \end{pmatrix}$	$\begin{pmatrix} 0 & 0 & 0 \\ 0 & 0 & 0 \\ 0 & 0 & 0 \end{pmatrix}$	0
G	$I_1 + J_1,$ $I_2 + J_2,$ $I_3 + J_3$	$\begin{pmatrix} a & 0 & 0 \\ 0 & a & 0 \\ 0 & 0 & a \end{pmatrix}$	$\begin{pmatrix} c & 0 & 0 \\ 0 & c & 0 \\ 0 & 0 & c \end{pmatrix}$	$\begin{pmatrix} a^2-c^2 & 0 & 0 \\ 0 & a^2-c^2 & 0 \\ 0 & 0 & a^2-c^2 \end{pmatrix}$	$\begin{pmatrix} 2ac & 0 & 0 \\ 0 & 2ac & 0 \\ 0 & 0 & 2ac \end{pmatrix}$	3

Table 3.4.2 Types of the Yang-Mills fields using the spinorial method and the corresponding classes using the little-group method.

Types of the fields using the spinorial method	The corresponding classes using the little group method
I_p	none
II_p	A, B, C_1, C_2
D_p	A, B, C_1, C_2
$III_p, IV_p, 0_p$	G
I_0, II_0	none
III_0	D, E, F
IV_0	D, E, F
0_0	D, E, F

Table 3.4.3 Classes of the Yang-Mills fields using the little-group method and the corresponding types using the spinorial one.

Classes of the fields using the little group method	The corresponding types using the spinorial method
A	II_p, D_p
B	II_p, D_p
C_1	II_p, D_p
C_2	II_p, D_p
D	$III_0, IV_0, 0_0$
E	$III_0, IV_0, 0_0$
F	$III_0, IV_0, 0_0$
G	$III_p, IV_p, 0_p$

4

Static Solutions of the Sourceless Yang-Mills Field Equations

In Chapters 2 and 3 we presented the Yang-Mills gauge field equations and also three different classification methods of the fields. In this chapter and in the following ones, we will solve these equations utilizing the classification schemes. The solutions considered by us are exact. Before we present these solutions, we will give a comparison of two fields. One of them, obtained by Carmeli, includes electric and magnetic parts, and the second was obtained by Morris. The comparison between these fields will be helpful to present the methods used in the sequel.

4.1 *Preliminaries*

The Morris field [62] describes static Yang-Mills systems and was obtained using the Kalb's ansatz [59], in the case of absence of external sources. This field has a spherical symmetry and describes a point magnetic monopole. We will compare [88] this field with the monopole field of Carmeli [60], which has both electric and magnetic charges, and give the conditions under which the two fields are identical. These fields provide a particular case to the t'Hooft-Polyakov ansatz. The Carmeli field was obtained using the Yang-Mills field equations written by means of the null tetrads (see Chapter 2). The Morris field was obtained by looking for time-independent solutions.

As has been mentioned by Morris, physical systems described by Yang-Mills fields with external sources admit magnetic monopole solutions. For example, the Georgi-Glashow model [27, 63], in which the gauge symmetry is spontaneously broken by the Higgs fields which are regarded as external sources for the Yang-Mills fields. One way to obtain solutions of the field equations is by using the t'Hooft-Polyakov ansatz. The solutions obtained by this way include those which describe magnetic monopoles without the appearance of the Dirac strings. Kalb has suggested an ansatz for time-independent Yang-Mills systems in the presence of external sources described by covariantly conserved currents and do not have to appear as Higgs fields. Kalb imposed a "London-type" constraint for obtaining soluble equations describing the Meissner effect. Morris [62] applied the Kalb ansatz to the case in which there are no external sources, and consequently obtained a magnetic monopole solution.

In Secs. 4.2 and 4.3, the two methods of solutions, those of Carmeli and Morris, are reviewed. In Sec. 4.4 we will compare these two solutions, whereas Sec. 4.5 is devoted to the summary.

4.2 *The Carmeli field*

In Chapter 2, we presented the formalism of writing the Yang-Mills field equations by means of the null tetrads, following Carmeli, Charach and Kaye [106]. In this formalism, in the flat space-time the field equations are given by Eqs. (2.2.27)–(2.2.30) and the relation between the potentials and the fields are given by Eqs. (2.2.23)–(2.2.25). Carmeli used these equations, where $J_{ab'k}$ was taken as zero, and obtained a solution that describes a monopole which has both electric and magnetic charges.

Analysis of the field variables shows that their angular dependence behaves like [60]:

$$\chi_0 \sim D^J_{1M}(\theta, \phi), \qquad \chi_1 \sim D^J_{0M}(\theta, \phi), \qquad \chi_2 \sim D^J_{-1M}(\theta, \phi), \qquad (4.2.1)$$

and

$$b_{00'} \sim D^J_{0M}(\theta, \phi), \qquad b_{01'} \sim D^J_{1M}(\theta, \phi),$$

$$b_{10'} \sim D^J_{-1M}(\theta, \phi), \qquad B_{11'} \sim D^J_{0M}(\theta, \phi),$$

(4.2.2)

where $D^J_{MN}(\theta, \phi)$ are the matrix elements of irreducible representations of the group SU(2) [123]. On the other hand, the isospin indices fix the second index of the matrix elements $D_{NM'}$. This leads to

$$\chi_{0, \pm 1} \sim D^J_{1, \pm 1}(\theta, \phi), \qquad \chi_{0,3} \sim D^J_{1,0}(\theta, \phi),$$

(4.2.3)

and the same for the rest of the other field variables. Here use has been made of the notation

$$\chi_{0, \pm 1} = \frac{1}{\sqrt{2}i}(\chi_{01} \mp \chi_{02}), \qquad \chi_{0,3} = \chi_{03}.$$

(4.2.4)

In this way the angular dependence of the field variables is determined.

For a monopole solution $J = 1$, and assuming a radial dependence of the potential in the form r^{-1}, one obtains [60]

$$b_{00'k} = \frac{2a}{gr} n_k(\theta, \phi),$$

(4.2.5)

$$b_{01'k} = \frac{ic}{g\sqrt{2r}} \mathcal{D}n_k(\theta, \phi),$$

(4.2.6)

$$b_{10'k} = -\frac{ic}{g\sqrt{2r}} \bar{\mathcal{D}}n_k(\theta, \phi),$$

(4.2.7)

$$b_{11'k} = \frac{b}{gr} n_k(\theta, \phi),$$

(4.2.8)

where a, b and c are arbitrary real constants, and $n_k(\theta, \phi)$ is a unit vector that is given by

$$n_k(\theta, \phi) = (\sin\theta\cos\phi, \sin\theta\sin\phi, \cos\theta), \tag{4.2.9}$$

along with the requirement

$$\chi_{0k} = \chi_{2k} = 0. \tag{4.2.10}$$

A simple calculation, using Eqs. (2.2.23)–(2.2.25), shows that the only non-vanishing component of the field is given by

$$\chi_{1k} = \frac{a + b + i}{2gr^2} n_k, \tag{4.2.11}$$

where c is taken as unity [60].

Now, one can show that the potentials given by Eqs. (4.2.5)–(4.2.8) and the field strengths given by Eqs. (4.2.10) and (4.2.11) provide a solution for the field Eqs. (2.2.27)–(2.2.30). Translating these results back into the standard notation, one finds that the potentials have the form

$$b_{0k} = \frac{a + b}{g} \frac{x^k}{r^2}, \tag{4.2.12}$$

$$b_{jk} = \frac{1}{g}\left\{-\varepsilon_{jki}\frac{x^i}{r^2} + (a - b)\frac{x^j x^k}{r^3}\right\}, \tag{4.2.13}$$

and the field strengths are given by

$$f_{0jk} = -\frac{a + b}{g} \frac{x^j x^k}{r^4}, \tag{4.2.14}$$

$$f_{ijk} = \frac{1}{g}\varepsilon_{ijn}\frac{x^n x^k}{r^4}, \tag{4.2.15}$$

where the coordinates are $x^0 = t$, $x^1 = x$, $x^2 = y$ and $x^3 = z$. The last results were obtained by using the following equations:

$$b_{\mu k} = n_\mu b_{00'k} - \bar{m}_\mu b_{01'k} - m_\mu b_{10'k} + l_\mu b_{11'k}, \tag{4.2.16}$$

$$f_{\mu\nu k} = \sigma_\mu^{ab'}\sigma_\nu^{cd'}f_{ab'cd'k} = \sigma_\mu^{ab'}\sigma_\nu^{cd'}(\varepsilon_{ac}\bar{\chi}_{b'd'} + \chi_{ack}\varepsilon_{b'd'}). \tag{4.2.17}$$

The Eqs. (4.2.16) and (4.2.17) are obtained from Eqs. (2.2.4), (2.2.5), (2.2.9) and (2.2.10) [60].

A direct substitution of the results (4.2.12)–(4.2.15) in the Yang-Mills field equations (2.1.11)–(2.1.14) shows that they constitute a solution to the sourceless field equations, that is $j^\mu = 0$.

4.3 *The Kalb Ansatz and the Morris field*

We now present the Morris solution of the Yang-Mills field equations [62]. The sourceless Yang-Mills field equations are given by (see Chapter 2)

$$\nabla^\nu \mathbf{f}_{\mu\nu} = 0, \tag{4.3.1}$$

$$\mathbf{f}_{\mu\nu} = \partial_\mu \mathbf{b}_\nu - \partial_\nu \mathbf{b}_\mu + \mathbf{b}_\mu \times \mathbf{b}_\nu, \tag{4.3.2}$$

where the coupling constant is taken as unity, and the covariant derivative is given by

$$\nabla^\mu = \partial^\mu + \mathbf{b}^\mu x. \tag{4.3.3}$$

[The fields defined by Eqs. (2.1.11) and (4.3.2) have opposite signs with respect to their potentials.] All the derivatives of the fields with respect to the time vanish. It can be shown that [62, 88]

$$\nabla^i \nabla_i \mathbf{b}_0 = 0, \tag{4.3.4}$$

$$\mathbf{b}_0 \times \nabla_i \mathbf{b}_0 = 0. \tag{4.3.5}$$

A solution for Eqs. (4.3.4) and (4.3.5) is given by

$$\nabla_i \mathbf{b}_0 = \alpha_i \mathbf{b}_0, \tag{4.3.6}$$

where α_i is a function of the spatial coordinates x^k. Consequently, one obtains

$$\partial^i \alpha_i + \alpha^i \alpha_i = 0, \tag{4.3.7}$$

for $\mathbf{b}_0 \neq 0$. Defining now the unit vector

$$\hat{\mathbf{b}}_0 = \mathbf{b}_0/b_0, \qquad b_0 = (\mathbf{b}_0 \cdot \mathbf{b}_0)^{1/2}, \tag{4.3.8}$$

and taking the inner product of Eq. (4.3.6) with the unit vector $\hat{\mathbf{b}}_0$, using Eq. (4.3.5), one obtains

$$\alpha_i = \partial_i \ln b_0. \tag{4.3.9}$$

Accordingly one obtains Laplace's equation for b_0

$$\partial^i \partial_i b_0 = 0. \tag{4.3.10}$$

Kalb presented the vector $\mathbf{b}_i$ in the form [59]

$$\mathbf{b}_i = -\hat{\mathbf{b}}_0 \times \partial_i \hat{\mathbf{b}}_0 + a_i \hat{\mathbf{b}}_0, \tag{4.3.11}$$

where

$$a_i = \hat{\mathbf{b}}_0 \cdot \mathbf{b}_i. \tag{4.3.12}$$

and the field strengths $\mathbf{f}_{ij}$ by

$$\mathbf{f}_{ij} = f_{ij}\hat{\mathbf{b}}_0, \tag{4.3.13}$$

where

$$f_{ij} = f_{ij}^{(A)} - \hat{\mathbf{b}}_0 \cdot (\partial_i \hat{\mathbf{b}}_0 \times \partial_j \hat{\mathbf{b}}_0), \tag{4.3.14}$$

and $f_{ij}^{(A)}$ is the Abelian part of the field strength,

$$f_{ij}^{(A)} = \partial_i a_j - \partial_j a_i. \tag{4.3.15}$$

Consequently, the magnetic and the electric field strengths are given by [62]

$$H^k = \tfrac{1}{2}\varepsilon^{kij} f_{ij}, \tag{4.3.16}$$

and

$$E_k = \hat{\mathbf{b}}_0 \cdot \mathbf{f}_{0k} = -\alpha_k b_0, \tag{4.3.17}$$

respectively, and α_i is given by Eq. (4.3.6).

From Eqs. (4.3.9), (4.3.10) and (4.3.17), one obtains

$$\partial^i E_i = 0, \tag{4.3.18}$$

and hence there are no electrical charges. However, the magnetic-charge density is given by

$$\partial_k H^k = -\tfrac{1}{2}\varepsilon^{kij}\partial^k \hat{\mathbf{b}}_0 \cdot (\partial_i \hat{\mathbf{b}}_0 \times \partial_j \hat{\mathbf{b}}_0), \tag{4.3.19}$$

and it does not vanish identically. Finally, one can obtain the a_i's from the following equation:

$$\partial^i f_{ij} = 0. \tag{4.3.20}$$

A solution to the Kalb ansatz can be obtained from Eq. (4.3.10). A general solution, with a radially directed field, is given by [62]

$$b_0^k = (\beta + \gamma/r)x^k/r, \tag{4.3.21}$$

where $r = (x^k x^k)^{1/2}$, β and γ are constants. Then one obtains the following relations:

$$(\partial_i \hat{\mathbf{b}}_0)^k = \frac{1}{r}\left(\delta_i^k - \frac{x^k x^i}{r^2}\right), \tag{4.3.22}$$

$$b_i^k = -\varepsilon^{kjn}\delta_i^n \frac{x^j}{r^2} + a_i \frac{x^k}{r}, \tag{4.3.23}$$

$$H^k = b^k - \frac{1}{2}\varepsilon^{kij}\varepsilon^{nlm}\delta_i^l \delta_j^m \frac{x^n}{r^3}, \tag{4.3.24}$$

where

$$b^k = \tfrac{1}{2}\varepsilon^{kij}f_{ij}^{(A)}, \tag{4.3.25}$$

and

$$\partial_k H^k = -4\pi\delta^3(\mathbf{x}).\tag{4.3.26}$$

4.4 *Comparison of the Carmeli and Morris fields*

We now compare the solutions obtained in Secs. 4.2 and 4.3, and show that each of them goes to the other one under certain conditions [88].

Starting with the Carmeli solution, given by Eqs. (4.2.12)–(4.2.15), obtained by the null-tetrad method. Taking $a = b = e/2$, one then gets

$$b_{0k} = \frac{ex^k}{gr^2},\tag{4.4.1}$$

$$b_{jk} = -\frac{1}{g}\,\varepsilon_{jkn}\frac{x^n}{r^2},\tag{4.4.2}$$

and

$$f_{0jk} = -\frac{ex^jx^k}{gr^4},\tag{4.4.3}$$

$$f_{ijk} = \frac{1}{g}\varepsilon_{ijn}\frac{x^nx^k}{r^4},\tag{4.4.4}$$

with the gauge condition $\partial_\mu b_k^\mu = 0$.

The new constant e can be interpreted as g times the electric charge of the monopole. Hence, the above solutions describe the field of a monopole that has both an electric charge e/g and a magnetic charge $1/g$. This can best be seen from the form of the field given by Eq. (4.2.11) when compared to the field of a monopole that has both electric and magnetic charges in electrodynamics. In this case, using the same method, one has for the Maxwell spinor [13]:

$$\phi_0 = \phi_2 = 0, \qquad \phi_1 = (e + iq)/2r^2.\tag{4.4.5}$$

Furthermore, if the electric charge e is taken to be zero in Eqs. (4.4.1)–(4.4.4), then one obtains the magnetic monopole field of Wu and Yang [21].

Now if we look upon the solution (4.3.21) and (4.3.23) obtained by Morris, and set $\beta = a_i = 0$, the solution reduces to

$$b_0^k = \gamma x^k / r^2, \tag{4.4.6}$$

$$b_k^i = -\varepsilon_{kij} x^j / r^2. \tag{4.4.7}$$

The above solution is a particular case of the Georgi-Glashow model with spontaneously broken SO(3) gauge symmetry in the limit of vanishing Higgs fields. To see this one writes the equations for the t'Hooft-Polyakov ansatz,

$$\phi_k(\mathbf{x}) = H(\xi) x^k / r^2, \tag{4.4.8}$$

$$W_k^0(\mathbf{x}) = J(\xi) x^k / r^2, \tag{4.4.9}$$

$$W_k^i(\mathbf{x}) = -\varepsilon_{kij}[1 - K(\xi)] x^j / r^2, \tag{4.4.10}$$

where $\boldsymbol{\phi}$ is the Higgs triplet, $\xi = ar$, and a is the vacuum value of one of the Higgs fields. Then the equations of motion are given by

$$\xi^2 \frac{d^2 K}{d\xi^2} = K(H^2 - J^2) + K(K^2 - 1), \tag{4.4.11}$$

$$\xi^2 \frac{d^2 H}{d\xi^2} = 2K^2 H + 2\lambda(H^2 - \xi^2)H, \tag{4.4.12}$$

$$\xi^2 \frac{d^2 J}{d\xi^2} = 2K^2 J. \tag{4.4.13}$$

By putting $\beta = 0$ in Eq. (4.3.21), $J(\xi) = \gamma$ and $H = 0$, one obtains $K = 0$, and then $\mathbf{W}_\mu$ coincides with $\mathbf{b}_\mu$. The condition $\beta = 0$ is a necessary condition for $\mathbf{b}_\mu = \mathbf{W}_\mu$. This solution describes fields with a point magnetic monopole without the explicit appearance of Dirac strings. The removal of the Higgs fields from the Georgi-Glashow model in the above form brings back the SO(3) symmetry.

Finally, taking $\gamma = -e$ in Eq. (4.4.6), one sees that the potentials in Eqs. (4.4.6) and (4.4.7) are the same as those given by Eqs. (4.4.1) and (4.4.2).

(Remember that the fields defined in the two methods have opposite signs.) As mentioned before, the two solutions were obtained under different assumptions. In the null-tetrad method, the angular dependence was taken like $D^1_{1M}(\theta, \phi)$ and the radial dependence as r^{-1} for the potentials $b^k_{ab'}$. The solution obtained under these conditions is then necessarily time independent and describes a field of monopole that has both electric and magnetic charges. On the other hand, the solution of Morris was obtained under the assumption that it is time independent and radially directed.

4.5 *Summary*

Two methods for obtaining static solutions of the sourceless Yang-Mills equations were presented. The first was the null-tetrad method given by Carmeli and the second was that of Morris which assumes time-independent solutions using the Kalb's ansatz. We showed that under certain conditions the two solutions coincide and reduce to a particular case of the t'Hooft-Polyakov ansatz in the limit of vanishing Higgs fields.

In the next chapter we classify some gauge fields.

5

Classification of Gauge Fields—Application

In this chapter, we will use one of the classification schemes, presented in Chapter 3, to classify some gauge fields that were obtained by Altamirano and Villarroel [109]. The scheme that we use is the Carmeli classification which is based on the distinct eigenvalues and eigenspinors of the gauge fields (see Sec. 3.1). This classification provides an example for the application of the classification scheme techniques [89].

5.1 *Preliminaries*

The gauge fields which we classify in this section are the solutions of the following two equations [109]:

$$\nabla^\mu f_{\mu\nu k} + \varepsilon_{kij} b_i^\mu f_{\mu\nu j} = 0, \tag{5.1.1}$$

$$\nabla^\mu {}^* f_{\mu\nu k} + \varepsilon_{kij} b_i^\mu {}^* f_{\mu\nu j} = 0. \tag{5.1.2}$$

The first of these equations is one of the standard Yang-Mills equations, and the second is the Bianchi identity for the Yang-Mills fields. No assumption is made of the usual relation between the Yang-Mills fields and potentials.

Analyses of the above equations were recently made by Altamirano and Villarroel [109], and a family of solutions was found by them. Although these solutions are not those of the Yang-Mills equations, the analysis of these two equations gives a better understanding to the solutions of the Yang-Mills equations.

5.2 The gauge fields in terms of the null tetrads

Altamirano and Villarroel [109] used the null-tetrad method of Carmeli, Charach and Kaye [60, 106] and of Newman [107, 108] to present the fields $f_{\mu\nu k}$ in the following form:

$$f_{\mu\nu k} = \tfrac{1}{2}[L_{\mu\nu}(\phi_k + \bar{\phi}_k) + \overline{M}_{\mu\nu}\chi_k + M_{\mu\nu}\bar{\chi}_k + N_{\mu\nu}\psi_k + \overline{N}_{\mu\nu}\bar{\psi}_k$$

$$+ P_{\mu\nu}(\phi_k - \bar{\phi}_k)], \tag{5.2.1}$$

where ϕ_k, χ_k and ψ_k are complex vectors in the isospin space, $L_{\mu\nu}$, $M_{\mu\nu}$, $N_{\mu\nu}$ and $P_{\mu\nu}$ are given by

$$L_{\mu\nu} = l_\mu n_\nu - l_\nu n_\mu, \tag{5.2.2}$$

$$M_{\mu\nu} = l_\mu m_\nu - l_\nu m_\mu, \tag{5.2.3}$$

$$N_{\mu\nu} = n_\mu m_\nu - n_\nu m_\mu, \tag{5.2.4}$$

$$P_{\mu\nu} = m_\mu \bar{m}_\nu - m_\nu \bar{m}_\mu. \tag{5.2.5}$$

Here, the null vectors l_μ, n_μ and m_μ are given by

$$l_\mu = \delta_\mu^0, \tag{5.2.6}$$

$$n_\mu = \delta_\mu^0 + \delta_\mu^1, \tag{5.2.7}$$

$$m_\mu = -\frac{r}{2}(\delta_\mu^2 + i \sin\theta\, \delta_\mu^3), \tag{5.2.8}$$

and $\bar{m}_\mu$ is the complex conjugate of m_μ. The coordinates here are u, r, θ and ϕ which are related to the Cartesian coordinates by

$$t = (1/\sqrt{2})(2u + r),$$

$$x = (1/\sqrt{2})r \sin \theta \cos \phi,$$

$$y = (1/\sqrt{2})r \sin \theta \sin \phi,$$

$$z = (1/\sqrt{2})r \cos \theta$$

(the coordinates in this section are multiplied by the factor $1/\sqrt{2}$ with respect to those in Chapter 2). Altamirano and Villarroel solved Eqs. (5.1.1) and (5.1.2) for the following special case:

$$\phi_3 = \phi \neq 0, \qquad \chi_2 = \chi \neq 0, \qquad \psi_1 = \psi \neq 0, \qquad (5.2.9)$$

(all other components vanish).

5.3 *Classification of Altamirano and Villarroel's fields*

To classify these fields, we use the Carmeli spinorial classification (see Sec. 3.1). Firstly, we look for the relations between the isovectors ϕ_k, χ_k, ψ_k and the symmetrical spinor χ_{ABk} which was defined by Eq. (3.1.23). From this equation and Eq. (5.2.9) one obtains the following:

$$\chi_{00k} = -\tfrac{1}{2}\bar{\psi}_k, \qquad \chi_{10k} = -\tfrac{1}{2}\bar{\phi}_k, \qquad \chi_{11k} = \tfrac{1}{2}\bar{\chi}_k. \qquad (5.3.1)$$

Using the relations (5.3.1) and the Eqs. (3.1.27), (3.1.30) and (3.1.31), defining the invariants P, G and H, respectively, one gets

$$P = -\frac{1}{2}\bar{\phi}^2, \qquad (5.3.2)$$

$$G = \frac{1}{24}(3\bar{\chi}^2\bar{\psi}^2 + 4\bar{\phi}^4), \qquad (5.3.3)$$

$$H = \frac{1}{144}(9\bar{\chi}^2\bar{\psi}^2 - 4\bar{\phi}^4)\bar{\phi}^2. \qquad (5.3.4)$$

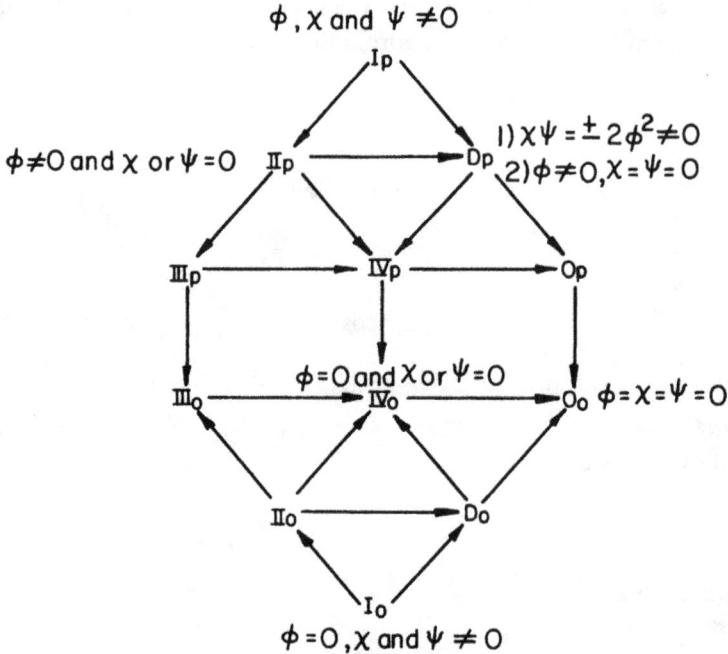

Fig. 5.3.1 Classification of Altamirano and Villarroel's fields. For the particular solutions for which ϕ, ψ, and χ do not vanish one obtains two different Carmeli classes: I_p and D_p. However, when one or more of the fields ϕ, ψ, χ vanish one obtains five different Carmeli classes: II_p, D_p, I_0, IV_0, and O_0. (The above-mentioned fields cannot be included in classes III_p, IV_p, O_p, II_0, D_0 and III_0.)

Now we classify the fields by means of the functions ϕ, χ and ψ instead of the invariants P, G and H which are given in terms of these functions (see Fig. 5.3.1).

In this classification we consider two cases:

(1) ϕ, χ, and ψ do not vanish (see Table 5.3.1). This is the case of the Altamirano and Villarroel solutions. They are included in two different Carmeli types (see Fig. 3.1.2):

(a) Type I_p, where ϕ, χ, and ψ can have arbitrary values different from zero ($P \neq 0$, $G^3 \neq 6H^2$).

(b) Type D_p, where $\chi\psi = \pm 2\phi^2 \neq 0$ ($P \neq 0$, $G^3 = 6H^2 \neq 0$).

(2) At least one of ϕ, χ and ψ is equal to zero (see Table 5.3.2). These fields are not included in the Altamirano and Villarroel solutions. They may be

Table 5.3.1 The Altamirano and Villarroel solutions in terms of the distinct numbers of the eigenspinors and the eigenvalues and their Carmeli's types.

Case	Values of ϕ, ψ, χ	Distinct eigenvalues	Distinct eigenspinors $\phi_{AB} = (\phi_{00}, \phi_{01}, \phi_{11})$	Invariants	The Carmeli type
a	$\phi, \psi, \chi \neq 0$	$\lambda_1' = -\frac{1}{3}\phi^2$ $\lambda_2' = \frac{1}{6}\phi^2 + \frac{1}{4}\psi\chi$ $\lambda_3' = \frac{1}{6}\phi^2 - \frac{1}{4}\psi\chi$	$(0,1,0)$ $(\bar\psi/\bar\chi, 0, 1)$ $(-\bar\psi/\bar\chi, 0, 1)$	$P \neq 0,\ G^3 \neq 6H^2$	I_p
b_1	$\psi\chi = -2\phi^2 \neq 0$	$\lambda_1' = \lambda_2' = -\frac{1}{3}\phi^2$ $\lambda_3' = \frac{2}{3}\phi^2$	$(0,1,0)$ $(\bar\psi/\bar\chi, 0, 1)$ $(-\bar\psi/\bar\chi, 0, 1)$	$P \neq 0,\ G^3 = 6H^2 \neq 0$	D_p
b_2	$\psi\chi = 2\phi^2 \neq 0$	$\lambda_1' = \lambda_3' = -\frac{1}{3}\phi^2$ $\lambda_2' = \frac{2}{3}\phi^2$	$(0,1,0)$ $(-\bar\psi/\bar\chi, 0, 1)$ $(\bar\psi/\bar\chi, 0, 1)$	$P \neq 0,\ G^3 = 6H^2 \neq 0$	D_p

Table 5.3.2 Particular cases of the Altamirano and Villarroel fields in terms of the distinct numbers of the eigenspinors and the eigenvalues and their Carmeli's types.

Case	Values of ϕ, ψ, χ	Distinct eigenvalues	Distinct eigenspinors $\phi_{AB} = (\phi_{00}, \phi_{01}, \phi_{11})$	Invariants	The Carmeli type
a_1	$\phi\,\chi \neq 0,\ \psi = 0$	$\lambda'_1 = -\tfrac{1}{3}\bar{\phi}^2$ $\lambda'_2 = \lambda'_3 = \tfrac{1}{6}\bar{\phi}^2$	$(0,1,0)$ $(0,0,1)$	$P \neq 0,\ G^3 = 6H^2 \neq 0$	II_p
a_2	$\phi, \psi \neq 0,\ \chi = 0$	$\lambda'_1 = -\tfrac{1}{3}\bar{\phi}^2$ $\lambda'_2 = \lambda'_3 = \tfrac{1}{6}\bar{\phi}^2$	$(0,1,0)$ $(1,0,0)$	$P \neq 0,\ G^3 = 6H^2 \neq 0$	II_p
b	$\phi \neq 0,\ \chi = \psi = 0$	$\lambda'_1 = -\tfrac{1}{3}\bar{\phi}^2$ $\lambda'_2 = \lambda'_3 = \tfrac{1}{6}\bar{\phi}^2$	$(0,1,0)$ $(1,0,0)$ $(0,1,0)$	$P \neq 0,\ G^3 = 6H^2 \neq 0$	D_p
c	$\phi = 0,\ \psi,\ \chi \neq 0$	$\lambda'_1 = 0$ $\lambda'_2 = \tfrac{1}{4}\sqrt{\bar{\psi}\bar{\chi}}$ $\lambda'_3 = -\tfrac{1}{4}\sqrt{\bar{\psi}\bar{\chi}}$	$(0,1,0)$ $(\bar{\psi}/\bar{\chi}, 0, 1)$ $(-\bar{\psi}/\bar{\chi}, 0, 1)$	$P = 0,\ G^3 \neq 6H^2$	I_0
d_1	$\phi = \psi = 0,\ \chi \neq 0$	$\lambda'_1 = \lambda'_2 = \lambda'_3 = 0$	$(0,1,0)$ $(0,0,1)$	$P = G = H = 0$	IV_0
d_2	$\phi = \chi = 0,\ \psi \neq 0$	$\lambda'_1 = \lambda'_2 = \lambda'_3 = 0$	$(1,0,0)$ $(0,1,0)$	$P = G = H = 0$	IV_0
e	$\phi = \chi = \psi = 0$	$\lambda'_1 = \lambda'_2 = \lambda'_3 = 0$	$(1,0,0)$ $(0,1,0)$ $(0,0,1)$	$P = G = H = 0$	O_0

classified into five different types according to the Carmeli classification scheme (see Fig. 3.1.2):

(a) II_p, $\phi \neq 0$ and ψ or $\chi = 0$ $(P \neq 0, G^3 = 6H^2 \neq 0)$.

(b) D_p, $\phi \neq 0$ and $\psi = \chi = 0$ $(P \neq 0, G^3 = 6H^2 \neq 0)$.

(c) I_0, where $\phi = 0$, ψ and $\chi \neq 0$ $(P = 0, G^3 \neq 6H^2)$.

(d) IV_0, $\phi = 0$ and ψ or $\chi = 0$ $(P = G = H = 0)$.

(e) O_0, $\phi = \psi = \chi = 0$ $(P = G = H = 0)$.

There are two cases where the considered fields cannot be included:

(1) When $P \neq 0$, $G = H = 0$ $(III_p, IV_p$ and $O_p)$, because if $P \neq 0$ then $\phi \neq 0$ by Eq. (5.3.2), $G = 0$ then $-3\psi^2\chi^2 = 4\phi^4$ by Eq. (5.3.3), and $H = 0$ then $9\psi^2\chi^2 = 4\phi^4$ by Eq. (5.3.4). From these two equations one gets -3 equal 9, which is wrong.

(2) When $P = 0$, $G^3 = 6H^2 \neq 0$ (II_0, D_0). $P = 0$ then $\phi = 0$ and so from Eq. (5.3.4) also $H = 0$, which contradicts the basic assumption that $H \neq 0$.

5.4 Summary

We saw that the Altamirano and Villarroel gauge-field solutions are of Carmeli's types I_p and D_p. Particular cases were subsequently shown to be of types II_p, D_p, I_0, IV_0, and O_0.

In the next chapter some exact solutions to the Yang-Mills field equations are given.

6

Exact Solutions of the Yang-Mills Field Equations

Gauge theory has an important role in particle physics, and hence there is a great interest in exact solutions of the Yang-Mills field equations (see, for example, Refs. 30–87). There is also a belief that classical solutions contribute greatly to the understanding of some topics in quantum theory.

6.1 *Preliminaries*

In this chapter, we will present exact solutions of the classical Yang-Mills field equations [90, 91, 93]. These solutions are obtained utilizing the classification of the fields according to known methods. Also, we use the Yang-Mills field equations that are written by means of the null tetrad (see Chapter 2), and consequently, we obtain a simple form to the field equations. We separate these solutions into two cases. In the first case, the fields belong to class A of the Castillejo, Kugler and Roskies classification scheme, and in the second one, they belong to class C_1 of the same classification scheme (see Sec. 3.4). In the following we will consider each case separately. It follows that for the first case one obtains a solution that has only a magnetic part (the electrical part vanishes), and it belongs to type D_p of the Carmeli classification, and to subcase 3_b of the Wang-Yang scheme (see Chapter 3).

In Sec. 6.2, we present the magnetic solution, whereas its classification is brought in Sec. 6.3. Exact solutions of the Yang-Mills field equations, which belong to class C_1, are subsequently given in Sec. 6.4. Sec. 6.5 is devoted to the summary.

6.2 *The magnetic solution*

The sourceless Yang-Mills field equations, in the special case where the fields belong to class A (see Table 3.4.1), and are written by means of the null tetrads, are given by [90, 91] (see Appendix A):

$$\left(\frac{\partial}{\partial u} - \frac{1}{2}\frac{\partial}{\partial r}\right)\beta_j = g\varepsilon_{kmj}\beta_k b_{11'm}, \tag{6.2.1}$$

$$\frac{\partial}{\partial r}\beta_j = g\varepsilon_{kmj}\beta_k b_{00'm}, \tag{6.2.2}$$

$$\frac{1}{\sqrt{2}r}\mathscr{D}\beta_j = g\varepsilon_{kmj}\beta_k b_{01'm}, \tag{6.2.3}$$

$$\frac{1}{\sqrt{2}r}\bar{\mathscr{D}}\beta_j = g\varepsilon_{kmj}\beta_k b_{10'm}, \tag{6.2.4}$$

(see Sec. 2.2). In these equations, β_k is a vector in the isospin space given by

$$\beta_k = (a + ib, ic, 0), \tag{6.2.5}$$

where a, b and c are real functions of the coordinates and describe the field strengths $f_{\mu\nu}^k$,

$$a(x) = f_{10}^1 = -f_{01}^1, \tag{6.2.6}$$

$$b(x) = f_{23}^1 = -f_{32}^1, \tag{6.2.7}$$

$$c(x) = f_{23}^2 = -f_{32}^2. \tag{6.2.8.}$$

The other components of the field strengths vanish.

From Eqs. (6.2.1)–(6.2.4) and the fact that the functions a, b, c, $b_{00'k}$ and $b_{11'k}$ are real, it follows that either a or $b_{00'k}$ and $b_{11'k}$ vanish. In the following, we will assume that $a = 0$. Consequently, one obtains from Eqs. (6.2.1)–(6.2.4) the following values for the potentials.

$$b_{00'3} = \frac{1}{gc} \frac{\partial b}{\partial r}, \tag{6.2.9}$$

$$b_{01'3} = \frac{1}{\sqrt{2grc}} \mathcal{D}b, \tag{6.2.10}$$

$$b_{11'3} = \frac{1}{gc} \left(\frac{\partial}{\partial u} - \frac{1}{2} \frac{\partial}{\partial r} \right) b, \tag{6.2.11}$$

$$b_{ab'1} = \frac{b}{c} b_{ab'2}, \tag{6.2.12}$$

along with the condition that $b^2 + c^2$ is a constant.

A particular solution of the above equations and conditions is seen to be given by

$$b(x) = K \cos \omega(\theta, \phi), \tag{6.2.13}$$

$$c(x) = K \sin \omega(\theta, \phi), \tag{6.2.14}$$

where K is a real constant and ω is a real function of θ and ϕ. Then, the field strengths are given by [see Eqs. (2.2.4) and (2.2.10)]

$$\chi_{abk} = \frac{1}{g} \chi_{ab} \alpha_k, \tag{6.2.15}$$

where χ_{ab} are given by

$$\chi_{00} = -\frac{i}{\sqrt{2}} (\cos \theta \cos \phi - i \sin \phi), \tag{6.2.16}$$

$$\chi_{01} = \chi_{10} = \frac{i}{2}\sin\theta\cos\phi, \tag{6.2.17}$$

$$\chi_{11} = \frac{1}{2}\bar{\chi}_{00}, \tag{6.2.18}$$

and α_k is a vector in the isospin space given by

$$\alpha_k = K[\cos\omega(\theta,\phi), \quad \sin\omega(\theta,\phi), \quad 0]. \tag{6.2.19}$$

From Eqs. (6.2.9)–(6.2.14) one obtains

$$b_{00'3} = b_{11'3} = 0, \tag{6.2.20}$$

$$b_{01'3} = -\frac{1}{\sqrt{2}gr}\mathcal{D}\omega(\theta,\phi). \tag{6.2.21}$$

It is easy to see that a general solution is given by

$$b(x) = K\cos\omega(u, r, \theta, \phi), \tag{6.2.22}$$

$$c(x) = K\sin\omega(u, r, \theta, \phi), \tag{6.2.23}$$

and thus

$$\alpha_k = K[\cos\omega(u, r, \theta, \phi), \sin\omega(u, r, \theta, \phi), 0]. \tag{6.2.24}$$

The choice of the particular solution given by Eqs. (6.2.13) and (6.2.14) is made in order to simplify the calculation of the potentials.

Now, from Eqs. (6.2.15)–(6.2.18), one obtains

$$\bar{\chi}_{00k} - 2\chi_{11k} = 0. \tag{6.2.25}$$

Then, from the relations between χ_{abk} and $b_{ab'k}$, Eqs. (2.2.23)–(2.2.25), we get the following equation:

$$\frac{1}{\sqrt{2}r}\bar{\mathcal{D}}(b_{00'k} + 2b_{11'k}) - 2\frac{\partial}{\partial u}b_{10'k} + g\varepsilon_{mjk}(b_{00'j} + 2b_{11'j})b_{10'm} = 0. \tag{6.2.26}$$

Now, assuming that $b_{01'k}$ does not depend on u, and using Eqs. (6.2.20) and (6.2.21), we obtain

$$b_{00'1} + 2b_{11'1} = \frac{1}{g} e^{f(u,r)} \cos \omega(\theta, \phi), \tag{6.2.27}$$

$$b_{00'2} + 2b_{11'2} = \frac{1}{g} e^{f(u,r)} \sin \omega(\theta, \phi), \tag{6.2.28}$$

where $f(u, r)$ is an arbitrary real function of u and r. Taking the sum of $\bar{\chi}_{00k}$ and $2\chi_{11k}$, one then finds

$$\sqrt{2}\frac{i}{g}(\cos\theta\cos\phi + i\sin\phi)\alpha_k = \frac{1}{\sqrt{2r}}\mathcal{D}(b_{00'k} - 2b_{11'k}) - 2\left(\frac{\partial}{\partial r} + \frac{1}{r}\right)b_{10'k}$$

$$+ g\varepsilon_{mjk}b_{10'm}(b_{00'j} - 2b_{11'j}). \tag{6.2.29}$$

If now one assumes that $b_{00'k} = 2b_{11'k}$, the latter equation then yields

$$\left(\frac{\partial}{\partial r} + \frac{1}{r}\right)b_{01'k} = \frac{i}{\sqrt{2g}}(\cos\theta\cos\phi - i\sin\phi)\alpha_k. \tag{6.2.30}$$

The solution of the last equation is given by

$$b_{01'k} = \frac{ir}{2\sqrt{2g}}(\cos\theta\cos\phi - i\sin\phi)\alpha_k + \frac{1}{\sqrt{2g}r}h_k(\theta, \phi), \tag{6.2.31}$$

where h_k are complex functions of θ and ϕ satisfying

$$h_1(\theta, \phi) = \cot\omega(\theta, \phi)h_2(\theta, \phi), \tag{6.2.32}$$

$$h_3(\theta, \phi) = -\mathcal{D}\omega. \tag{6.2.33}$$

From the real part of χ_{01k} one obtains the following condition on the function $f(u, r)$:

$$\left(\frac{\partial}{\partial u} - \frac{\partial}{\partial r}\right) f(u, r) = 0, \tag{6.2.34}$$

from which one finds that $f(u, r) = f(u + r)$. From the imaginary part of χ_{01k} one obtains

$$I_m[\mathscr{D} + \cot \theta - \mathscr{D}(\ln \cos \omega)]\bar{h}_1(\theta, \phi) = 0. \tag{6.2.35}$$

Summarizing the above results, we get the following for the potentials:

$$b_{00'k} = \frac{1}{g} e^{f(u+r)} \alpha_k, \tag{6.2.36}$$

$$b_{01'k} = \frac{ir}{2\sqrt{2g}} (\cos \theta \cos \phi - i \sin \phi)\alpha_k + \frac{1}{\sqrt{2gr}} h_k(\theta, \phi), \tag{6.2.37}$$

$$b_{11'k} = \frac{1}{2g} e^{f(u+r)} \alpha_k, \tag{6.2.38}$$

and $b_{10'k}$ is the complex conjugate of $b_{01'k}$.

The above potentials, along with the field strengths given by Eqs. (6.2.15)–(6.2.18), constitute a solution of the Yang–Mills field equations.

In the Minkowskian coordinates, the above solution gets the following form:

$$b_\mu^k = \frac{1}{g} (A_\mu \alpha^k + B_\mu h_{\text{Re}}^k + D_\mu h_{\text{Im}}^k), \tag{6.2.39}$$

$$f_{\mu\nu}^k = \frac{1}{g} f_{\mu\nu} \alpha^k, \tag{6.2.40}$$

where

$$A_\mu = (e^{f(t)}, 0, \tfrac{1}{2}z, -\tfrac{1}{2}y), \tag{6.2.41}$$

$$B_\mu = \frac{1}{r^2 R} (0, xz, yz, -R^2), \tag{6.2.42}$$

$$D_\mu = \frac{1}{rR}(0, -y, x, 0),$$
(6.2.43)

$$f_{23} = -f_{32} = 1 \qquad \text{(other components vanish).}$$
(6.2.44)

In the latter equations, $R^2 = x^2 + y^2$, $r^2 = R^2 + z^2$, h_{Re}^k and h_{Im}^k are the real and the imaginary parts of $h^k(\theta, \phi)$, respectively.

This solution describes a unidirectional constant field in space-time, whose direction in the isospin space is determined by the vector α_k given by Eq. (6.2.19). Hence one should expect the energy density of the field to be a constant throughout the space-time. This can easily be seen if one calculates the energy density

$$T_{00} = \frac{1}{2\pi} \sigma_0^{ac'} \sigma_0^{bd'} \chi_{abk} \overline{\chi}_{c'd'k},$$
(6.2.45)

where $T_{\mu\nu}$ is the energy-momentum tensor (see Sec. 2.2). One then finds

$$T_{00} = \frac{K^2}{8\pi g^2}.$$
(6.2.46)

As has been mentioned above, using the null-tetrad method in gravitation brought a remarkable simplification in the Einstein field equations, and this enabled Kinnersley to find all type D vacuum solutions [105]. We see here that the extension of the null-tetrad method to the non-Abelian gauge theory also brought a significant simplification to the gauge field equations which enabled us to obtain the magnetic solution and the other solutions. This method was also used in obtaining the Carmeli monopole that has both electric and magnetic parts [60] (see Chapter 4), and other solutions obtained by Altamirano and Villarroel [109] (see Chapter 5). Other solutions were obtained, using other classification schemes, by Castillejo and Kugler [104].

6.3 *Classification of the magnetic solution*

Now we classify the solution, given in Sec. 6.2, according to the three classification schemes presented in Chapter 3.

6.3.1 *The classification by the Carmeli spinorial method* (see Sec. 3.1)

The invariants P, G and H given by Eqs. (3.1.27), (3.1.30) and (3.1.31), in this case will get the following values:

$$P = -\frac{2}{g^2}(\chi_{00}\chi_{11} - \chi_{10}^2) = \frac{1}{2}\frac{K^2}{2g^2}, \tag{6.3.1}$$

$$G = \frac{2}{3}P^2 = \frac{1}{6}\frac{K^4}{g^4}, \tag{6.3.2}$$

$$H = \frac{2}{9}P^3 = \frac{1}{36}\frac{K^6}{g^6}. \tag{6.3.3}$$

To get these results, we used Eqs. (3.1.24), (3.1.29), and Eqs. (6.2.15)–(6.2.18).

Consequently, the magnetic solution is characterized by invariants satisfying $P \neq 0$ and $G^3 = 6H^2 \neq 0$, and thus it belongs to either type II$_p$ or D$_p$ (see Fig. 3.1.2).

To determine whether this solution is of type II$_p$ or D$_p$ one has to find the eigenvalues and the eigenspinors of the Eq. (3.1.33). The eigenvalues can be found from Eq. (3.1.35), using the values of G and H as given by Eqs. (6.3.2) and (6.3.3). One then obtains

$$\lambda'^3 - \frac{K^4}{12g^4}\lambda' - \frac{K^6}{108g^6} = 0, \tag{6.3.4}$$

whose roots are easily found to be given by

$$\lambda_1' = \frac{1}{3}\frac{K^2}{g^2}, \tag{6.3.5}$$

$$\lambda_2' = \lambda_3' = \frac{1}{6}\frac{K^2}{g^2}. \tag{6.3.6}$$

The corresponding eigenspinors are also found to be given by

$$\boldsymbol{\phi}_1 = \left(1, -\frac{\chi_{10}}{\chi_{11}}, \frac{\chi_{00}}{\chi_{11}}\right), \tag{6.3.7}$$

$$\boldsymbol{\phi}_2 = \left(1, 0, -\frac{\chi_{00}}{\chi_{11}}\right), \tag{6.3.8}$$

$$\boldsymbol{\phi}_3 = \left(0, -1, \frac{2\chi_{10}}{\chi_{11}}\right), \tag{6.3.9}$$

where χ_{00}, χ_{10} and χ_{11} are given by Eqs. (6.2.16)–(6.2.18), and the three-component notation used in the last equations is defined by

$$\boldsymbol{\phi} = (\phi^{00}, \phi^{01} = \phi^{10}, \phi^{11}). \tag{6.3.10}$$

The eigenspinor $\boldsymbol{\phi}_1$ corresponds to the eigenvalue λ'_1, whereas the eigenspinors $\boldsymbol{\phi}_2$ and $\boldsymbol{\phi}_3$ correspond to λ'_2 and λ'_3, respectively. Consequently, there are two distinct eigenvalues and three distinct eigenspinors. Hence, the solution is of type D_p (see Fig. 3.1.2).

6.3.2 *The classification by the Castillejo-Kugler-Roskies little group method* (see Sec. 3.4)

As we mentioned in the beginning of this section, we started by a field that belongs to class A of this classification. The general form of this field (see Table 3.4.1), expressed by the electric and the magnetic parts, is given by

$$E = \begin{pmatrix} a & 0 & 0 \\ 0 & 0 & 0 \\ 0 & 0 & 0 \end{pmatrix}, \tag{6.3.11}$$

$$H = \begin{pmatrix} b & c & 0 \\ 0 & 0 & 0 \\ 0 & 0 & 0 \end{pmatrix}, \tag{6.3.12}$$

whereas the symmetrical matrices K_{ij} and J_{ij} are

$$K = \begin{pmatrix} a^2 - b^2 & -bc & 0 \\ -bc & -c^2 & 0 \\ 0 & 0 & 0 \end{pmatrix}, \tag{6.3.13}$$

$$J = \begin{pmatrix} 2ab & ac & 0 \\ ac & 0 & 0 \\ 0 & 0 & 0 \end{pmatrix} \tag{6.3.14}$$

In this case $a = 0$, b and c are given by Eqs. (6.2.13) and (6.2.14). (Our convention for the matrices E and H is in accordance to Wang and Yang which are the transposed to those given in Ref. 102.)

6.3.3 *The classification by the Wang-Yang matrix method* (see Sec. 3.2)

The classification by this method is based on the rank of the matrix Δ which is given by Eq. (3.2.9). In the case of the magnetic solution, it is given by

$$\Delta = \begin{pmatrix} (a + ib)^2 & i(a + ib)c & 0 \\ i(a + ib)c & -c^2 & 0 \\ 0 & 0 & 0 \end{pmatrix}. \tag{6.3.15}$$

It is obvious that this matrix can be realized by the matrix A in the following form:

$$A = \begin{pmatrix} a + ib & ic & 0 \\ 0 & 0 & 0 \\ 0 & 0 & 0 \end{pmatrix}, \tag{6.3.16}$$

such that $\Delta = A^t A$ (it should be remembered that the matrices that give E and H, of Wang and Yang, are the transposed of those of Castillejo, Kugler, and Roskies). In the case $a = 0$, Δ then becomes

$$\Delta = -\begin{pmatrix} b^2 & cb & 0 \\ bc & c^2 & 0 \\ 0 & 0 & 0 \end{pmatrix}, \tag{6.3.17}$$

which can also be presented as follows:

$$\Delta = -\begin{pmatrix} b \\ c \\ 0 \end{pmatrix} (b \quad c \quad 0), \tag{6.3.18}$$

where $b = K \cos \omega(\theta, \phi)$ and $c = K \sin \omega(\theta, \phi)$. One can easily see that the rank of the matrix Δ is 1, and hence the solution is of case 3 of the Wang and Yang classification method. Furthermore, the ratio $c/b = \tan \omega(\theta, \phi)$ is real, and consequently the solution is of subcase 3_b of this classification method.

In summary, we have presented a magnetic solution of the classical Yang-Mills field equations. The solution belongs to type D of the Carmeli classification. The field strengths (but not the potentials) have the Abelian form $f_{\mu\nu}^k = f_{\mu\nu} \alpha^k$. The solution was obtained using the null-tetrad method, and was classified according to three known classification schemes. In the next chapter we will give the equations of motion for a classical particle moving in this field.

In the following section we present more exact solutions of the Yang-Mills field equations which are obtained utilizing the same methods that were used to get the magnetic solution. As has been said, these solutions belong to class C_1 of the Castillejo-Kugler-Roskies classification.

6.4 *Exact solutions of class* C_1

The matrix $A_i^k = E_i^k + iH_i^k$, describing the Yang-Mills fields that belong to class C_1, is given by (see Table 3.4.1)

$$A_i^k = \begin{pmatrix} a + ic & 0 & 0 \\ 0 & b + id & 0 \\ 0 & 0 & b + id \end{pmatrix}, \tag{6.4.1}$$

where E_i^k and H_i^k are the electric and the magnetic parts of the Yang-Mills fields, respectively. The quantities a, b and c are arbitrary real functions of the coordinates.

In the null-tetrad method, the fields and the potentials are given by the spinors χ_{abk} and the isotriplets $b_{ac'k}$, respectively (see Chapter 2). In class C_1, the χ_{abk} take the following forms [93]:

$$\chi_{00k} = \frac{-1}{\sqrt{2}} [(a + ic)\mathscr{D}\gamma_k + (b + id)\mathscr{D}\delta_k], \tag{6.4.2}$$

$$\chi_{01k} = \frac{1}{2}[(a + ic)\gamma_k + (b + id)\delta_k],$$

(6.4.3)

$$\chi_{11k} = \frac{1}{2\sqrt{2}}[(a + ic)\bar{\mathscr{D}}\gamma_k + (b + id)\bar{\mathscr{D}}\delta_k].$$

(6.4.4)

Here

$$\gamma_k = (n_1, 0, 0),$$

(6.4.5)

$$\delta_k = (0, n_2, n_3),$$

(6.4.6)

and n_k is the unit vector given by

$$n_k = (\sin\theta\cos\phi, \sin\theta\sin\phi, \cos\theta).$$

(6.4.7)

If we use the values of χ_{abk} from Eqs. (6.4.2)–(6.4.4) in the field equations (2.2.27)–(2.2.30) when $j_{ab'k} = 0$, we obtain the following relations between the potentials (see Appendix B):

$$b_{00'2}n_2 + \sqrt{2}b_{10'2}\mathscr{D}n_2 = b_{00'3}n_3 + \sqrt{2}b_{10'3}\mathscr{D}n_3,$$

(6.4.8)

$$-b_{00'2}\bar{\mathscr{D}}n_2 + \sqrt{2}b_{10'2}n_2 = -b_{00'3}\bar{\mathscr{D}}n_3 + \sqrt{2}b_{10'3}n_3,$$

(6.4.9)

$$\sqrt{2}b_{11'2}n_2 - b_{01'2}\bar{\mathscr{D}}n_2 = \sqrt{2}b_{11'3}n_3 - b_{01'3}\bar{\mathscr{D}}n_3,$$

(6.4.10)

$$\sqrt{2}b_{11'2}\mathscr{D}n_2 + b_{01'2}n_2 = \sqrt{2}b_{11'3}\mathscr{D}n_3 + b_{01'3}n_3.$$

(6.4.11)

Now, one has to solve the system of the four algebraic equations (6.4.8)–(6.4.11) of complex variables. One then obtains two independent solutions leading to the two cases:

Case 1:

$$b_{00'2} = -2b_{11'2} = -n_3 F,$$

(6.4.12)

$$b_{00'3} = -2b_{11'3} = n_2 F,$$

(6.4.13)

$$b_{01'2} = \frac{-1}{\sqrt{2}} F \mathscr{D} n_3, \tag{6.4.14}$$

$$b_{01'3} = \frac{1}{\sqrt{2}} F \mathscr{D} n_2. \tag{6.4.15}$$

Case 2:

$$b_{00'2} = -2b_{11'2} = n_2 H, \tag{6.4.16}$$

$$b_{00'3} = -2b_{11'3} = n_3 H, \tag{6.4.17}$$

$$b_{01'2} = \frac{1}{\sqrt{2}} H \mathscr{D} n_2, \tag{6.4.18}$$

$$b_{01'3} = \frac{1}{\sqrt{2}} H \mathscr{D} n_3. \tag{6.4.19}$$

Here F and H are real functions of the coordinates u, r, θ and ϕ. Now to complete the solution, it remains to find the functions F, H, a, b, c, d and the potentials $b_{00'1}$, $b_{01'1}$ and $b_{11'1}$.

Let us first consider the Case 1. Substituting the potentials (6.4.12)–(6.4.15) in the field equations (2.2.27)–(2.2.30), where $j_{ab'k} = 0$, and the relations between the fields and potentials (2.2.23)–(2.2.25), then one finds that all the field variables are independent of u and that the following relations exist (see Appendix C):

$$n_1(b_{00'1} - 2b_{11'1}) = -\sqrt{2}(b_{01'1} \bar{\mathscr{D}} n_1 + b_{10'1} \mathscr{D} n_1), \tag{6.4.20}$$

$$\frac{i}{grF}(\mathscr{D} n_1 \bar{\mathscr{D}} F - \bar{\mathscr{D}} n_1 \mathscr{D} F) = b_{00'1} - 2b_{11'1}, \tag{6.4.21}$$

$$b = -\tfrac{1}{2} g F(b_{00'1} + 2b_{11'1}), \tag{6.4.22}$$

$$d = \frac{1}{n_1} \left[\frac{ig}{\sqrt{2}} F(b_{01'1} \bar{\mathscr{D}} n_1 - b_{10'1} \mathscr{D} n_1) - \frac{\partial F}{\partial r} \right], \tag{6.4.23}$$

$$c\mathscr{D}n_1 = \frac{i}{2r}\mathscr{D}(b_{00'1} - 2b_{11'1})$$

$$- i\sqrt{2}\left(\frac{\partial}{\partial r} + \frac{1}{r}\right)b_{01'1} - gF^2\mathscr{D}n_1, \tag{6.4.24}$$

$$a\mathscr{D}n_1 = \frac{-1}{2r}\mathscr{D}(b_{00'1} + 2b_{11'1}), \tag{6.4.25}$$

$$an_1 = -\frac{1}{2}\frac{\partial}{\partial r}(b_{00'1} + 2b_{11'1}). \tag{6.4.26}$$

From the last two equations one obtains

$$\frac{n_1}{r}\mathscr{D}(b_{00'1} + 2b_{11'1}) = \mathscr{D}n_1\frac{\partial}{\partial r}(b_{00'1} + 2b_{11'1}), \tag{6.4.27}$$

whose solution is given by

$$b_{00'1} + 2b_{11'1} = 2M(x), \tag{6.4.28}$$

where $M(x)$ is a real function of the variable $x = rn_1 = r\sin\theta\cos\phi$. Using Eq. (6.4.28) in Eqs. (6.4.22), (6.4.25) and (6.4.26), one gets

$$a = -M', \tag{6.4.29}$$

$$b = -gFM, \tag{6.4.30}$$

where the prime denotes a derivative with respect to the variable x. Inserting the results (6.4.29) and (6.4.30) in the field equations (2.2.27)–(2.2.30); one finds that F is a function of x only, and hence

$$b_{00'1} = 2b_{11'1} = M(x), \tag{6.4.31}$$

$$b_{01'1} = 0, \tag{6.4.32}$$

$$c = -gF^2, \tag{6.4.33}$$

$$d = -F'. \tag{6.4.34}$$

Furthermore, the functions F and M have to satisfy the following differential equations:

$$M'' - 2g^2 F^2 M = 0, \tag{6.4.35}$$

$$F'' - g^2 F(F^2 - M^2) = 0. \tag{6.4.36}$$

In the same way, we get the following results for Case 2 (see Appendix D):

$$b_{00'1} = -2b_{11'1} = n_1 G(t), \tag{6.4.37}$$

$$b_{01'1} = \frac{1}{\sqrt{2}} \mathcal{D} n_1 G(t), \tag{6.4.38}$$

$$a = \dot{G}, \tag{6.4.39}$$

$$b = \dot{H}, \tag{6.4.40}$$

$$c = -gH^2, \tag{6.4.41}$$

$$d = -gHG, \tag{6.4.42}$$

where a dot on G and H denotes differentiation with respect to the coordinate $t = r + u$. Here G and H are real functions of t only satisfying the following differential equations:

$$\ddot{G} + 2g^2 H^2 G = 0, \tag{6.4.43}$$

$$\ddot{H} + g^2 H(H^2 + G^2) = 0. \tag{6.4.44}$$

To find the solution for each case, we have to solve the system of two ordinary nonlinear differential equations (6.4.35), (6.4.36) and (6.4.43), (6.4.44). Now we look for particular solutions for cases 1 and 2.

(1) Let us assume that the electrical part of the Yang-Mills field vanishes, and therefore $M = 0$. Then integrating Eq. (6.4.36) gives

$$F'^2 = (g^2 F^4 \pm A^4)/2, \tag{6.4.45}$$

where $\pm A^4/2$ is an integration constant. The integration of the last equation is an elliptic integral of the first kind [124], and from it one obtains for $+A^4$

$$F = \pm \frac{A}{\sqrt{g}} \sqrt{\frac{1 - cn\chi}{1 + cn\chi}} . \tag{6.4.46}$$

For $-A^4$ one obtains:

$$F = \pm \frac{A}{\sqrt{g}} \sqrt{\frac{1 + cn^2\psi}{1 - cn^2\psi}} , \tag{6.4.47}$$

$$F = \frac{A}{\sqrt{g} cn\psi} , \tag{6.4.48}$$

where $\chi = \sqrt{2g}A(x - x_0)$, $\psi = \chi/2$ and $cn\chi$ is a Jacobian elliptic function.

(2) In this case we take $H = G$, and then by integrating Eq. (6.4.44) one gets

$$\dot{H}^2 = B^4 - g^2 H^4 , \tag{6.4.49}$$

where B^4 is an integration constant. The integration of Eq. (6.4.49) is an elliptic integral of the first kind, from which one obtains

$$H = \pm \frac{B}{\sqrt{g}} \sqrt{\frac{1 - cn^2\tau}{1 + cn^2\tau}} , \tag{6.4.50}$$

$$H = \frac{B}{\sqrt{g}} cn\tau , \tag{6.4.51}$$

where $\tau = \sqrt{2g}B(t - t_0)$.

We give the Yang-Mills potentials and fields in their ordinary form in Cartesian coordinates:

Case 1:

$$b_\mu^k = (M\delta_1^k, 0, F\delta_3^k, F\delta_2^k), \tag{6.4.52}$$

$$f_{01}^k = \delta_1^k M', \tag{6.4.53}$$

$$f_{02}^k = g\delta_2^k FM, \tag{6.4.54}$$

$$f_{03}^k = -g\delta_3^k FM, \tag{6.4.55}$$

$$f_{12}^k = -\delta_3^k F', \tag{6.4.56}$$

$$f_{23}^k = g\delta_1^k F^2, \tag{6.4.57}$$

$$f_{31}^k = \delta_2^k F'. \tag{6.4.58}$$

Case 2:

$$b_\mu^k = (0, G\delta_1^k, H\delta_2^k, H\delta_3^k), \tag{6.4.59}$$

$$f_{01}^k = -\delta_1^k \dot{G}, \tag{6.4.60}$$

$$f_{02}^k = -\delta_2^k \dot{H}, \tag{6.4.61}$$

$$f_{03}^k = -\delta_3^k \dot{H}, \tag{6.4.62}$$

$$f_{12}^k = -g\delta_3^k GH, \tag{6.4.63}$$

$$f_{23}^k = -g\delta_1^k H^2, \tag{6.4.64}$$

$$f_{31}^k = -g\delta_2^k GH. \tag{6.4.65}$$

Finally, one can calculate the energy density T_{00}. Using the values of χ_{abk} given by Eqs. (6.4.2)–(6.4.4) in Eq. (6.2.45), we obtain

$$T_{00} = \frac{1}{8\pi}(a^2 + c^2 + 2b^2 + 2d^2). \tag{6.4.66}$$

In Case 1, inserting the results (6.4.29), (6.4.30), (6.4.33) and (6.4.34) in the last equation, we get

$$T_{00} = \frac{1}{8\pi}[M'^2 + 2F'^2 + g^2F^2(F^2 + 2M^2)]. \tag{6.4.67}$$

In Case 2, substituting the results (6.4.39)–(6.4.42) in Eq. (6.4.66), we obtain

$$T_{00} = \frac{1}{8\pi} [\dot{G}^2 + 2\dot{H}^2 + g^2 H^2 (H^2 + 2G^2)]. \qquad (6.4.68)$$

6.5 *Summary*

In this chapter we presented exact solutions of the classical Yang-Mills field equations. To get these solutions we started with fields that belong to some classes of known classification schemes, and therefore they have certain symmetry. This symmetry, together with the use of the Carmeli-Charach-Kaye null-tetrad formalism, simplified the calculation. The methods used above and the solutions obtained can be used as a basis to deriving other exact solutions of the Yang-Mills field equations, which is an important subject in particle physics and quantum theory.

We classified the first solution, which has only a magnetic part, according to different classification schemes, and the energy density was calculated in each case.

In the next chapter, we give the equations of motion of a test particle moving in Yang-Mills fields, and also give the solutions of these equations for particular cases, including the fields given in this chapter.

7

The Motion of a Test Particle in Classical Yang-Mills Fields

In the previous chapters we gave a review of gauge fields and their classification, and also presented the Yang-Mills field equations by means of the null tetrads. Subsequently, we used these methods to obtain and classify exact solutions of the Yang-Mills field equations.

In this chapter, for a better understanding of the physical meaning of these solutions, we discuss the equations of motion of a test particle moving in Yang-Mills fields [110, 113]. We solve [91, 92] these equations for two particular Yang-Mills fields, the SU(2) magnetic field [90, 91] given in the last chapter and the instanton field [67] which is of great interest in quantum theory.

7.1 *Preliminaries*

The problem of the motion of a test particle in classical Yang-Mills field has attracted a good deal of attention, and several works were published on this subject [110–116]. In this chapter, we present two methods for deriving the equations of motion of a particle in the Yang-Mills field. The first is that of Wong [110], in which the equations of motion of the particle are obtained from quantum theory by taking appropriate limits. The second is of Drechsler and Rosenblum [113], where the same equations of motion are derived from energy and momentum, and from covariant isotopic charge, conservation.

In Secs. 7.2 and 7.3, we review the derivation of the equations of motion of a particle moving in classical Yang-Mills field according to the approaches of Wong, Drechsler and Rosenblum, respectively. In Secs. 7.4 and 7.5, we present the solutions of the equations of motion of a test particle moving in the SU(2) magnetic and the instanton fields, respectively. Sec. 7.6 is devoted to the summary.

7.2 *The equations of motion according to Wong*

As has been mentioned, the equations of motion of a test particle moving in the classical Yang-Mills field were obtained by Wong from quantum field equations by taking appropriate limit. One starts from the Lagrangian density of the interaction of the Yang-Mills field and a particle with isotopic spin $\frac{1}{2}$, given by [110]

$$\mathcal{L} = -\frac{1}{4}\mathbf{f}^{\mu\nu}\cdot\mathbf{f}_{\mu\nu} - \bar{\psi}\gamma^{\mu}(\partial_{\mu} - ig\mathbf{b}_{\mu}\cdot\mathbf{X})\psi - \frac{mc}{\hbar}\bar{\psi}\psi, \qquad (7.2.1)$$

where ψ is the field of the particle, $\mathbf{X}$ are the generators of SU(2) satisfying the commutation relations

$$[X_i, X_j] = i\varepsilon_{ijk}X_k, \qquad (7.2.2)$$

and γ_{μ} are Hermitian matrices satisfying the anticommutation relations

$$\{\gamma_{\mu}, \gamma_{\nu}\} = 2\delta_{\mu\nu}. \qquad (7.2.3)$$

Now, the field equations are given by

$$\partial^{\nu}\mathbf{f}_{\mu\nu} + g\mathbf{b}^{\nu} \times \mathbf{f}_{\mu\nu} = -ig\bar{\psi}\gamma_{\mu}\mathbf{X}\psi, \qquad (7.2.4)$$

$$\gamma^{\mu}(\partial_{\mu} - ig\mathbf{X}\cdot\mathbf{b}_{\mu})\psi + \frac{mc}{\hbar}\psi = 0, \qquad (7.2.5)$$

where $\mathbf{f}_{\mu\nu}$ is given by Eq. (2.1.11),

$$\mathbf{f}_{\mu\nu} = \partial_\nu \mathbf{b}_\mu - \partial_\mu \mathbf{b}_\nu - g\mathbf{b}_\mu \times \mathbf{b}_\nu. \tag{7.2.6}$$

(Our **f** has opposite sign to that of Wong [110].)

Equation (7.2.5), considered as a one-particle Dirac equation describing a particle with isotopic spin in the external Yang-Mills field, can be written in the form of Schrödinger equation,

$$i\hbar \frac{\partial \psi}{\partial t} = H\psi, \tag{7.2.7}$$

with the Hamiltonian H given by

$$H = c\alpha_i(p_i - g\mathbf{b}_i \cdot \mathbf{I}) - imc^2\beta - gc\mathbf{b}_0 \cdot \mathbf{I}. \tag{7.2.8}$$

Here, α_i and β are the Dirac matrices, $p_i = (\hbar/i)\partial_i$, and $\mathbf{I} = \hbar\mathbf{X}$ satisfying the commutation relations

$$[I_i, I_j] = i\hbar\varepsilon_{ijk}I_k. \tag{7.2.9}$$

In the Heisenberg picture, one obtains the equations of motion

$$\frac{dx_i}{dt} = \frac{i}{\hbar}[H, x_i] = c\alpha_i, \tag{7.2.10}$$

$$\frac{dp_i}{dt} = \frac{i}{\hbar}[H, p_i] = gc(\alpha_j\partial_i\mathbf{b}_j + \partial_i\mathbf{b}_0) \cdot \mathbf{I}, \tag{7.2.11}$$

$$\frac{d\mathbf{I}}{dt} = \frac{i}{\hbar}[H, \mathbf{I}] = -g\left(\frac{dx_i}{dt}\mathbf{b}_i + c\mathbf{b}_0\right) \times \mathbf{I}. \tag{7.2.12}$$

Defining the mechanical momenta by

$$\pi_i = p_i - g\mathbf{b}_i \cdot \mathbf{I}, \tag{7.2.13}$$

one then gets

$$\frac{d\pi_i}{dt} = g\left(\frac{dx_j}{dt}\mathbf{f}_{ij} + c\mathbf{f}_{i0}\right) \cdot \mathbf{I}.$$

(7.2.14)

Comparing Eq. (7.2.14) with the corresponding equation of motion for a charged particle moving in an external electromagnetic field, the following equation describing the world line $\zeta^\mu(\tau)$ of the particle in space-time suggests itself:

$$m\ddot{\zeta}^\mu = g\mathbf{f}^{\mu\nu} \cdot \mathbf{I}\dot{\zeta}_\nu,$$

(7.2.15).

where the dot denotes differentiation with respect to the proper time τ. One can see that the right-hand side of Eq. (7.2.15) describes a generalization of the Lorentz force. In addition, one obtains

$$\dot{\mathbf{I}} + g\mathbf{b}^\mu \times \mathbf{I}\dot{\zeta}_\mu = 0.$$

(7.2.16)

Thus, the particle is described by an internal vector $\mathbf{I}$ as well as its space-time coordinates ζ^μ.

The isotopic current carried by a point particle, analogous to the electric current, is given by

$$g\int \mathbf{I}(\tau)\dot{\zeta}_\mu \delta^{(4)}[x - \zeta(\tau)]d\tau.$$

(7.2.17)

Hence the field equations will be those of Yang-Mills with the current $\mathbf{j}^\mu$ that is a sum of terms of the form (7.2.17). The Eqs. (7.2.15) and (7.2.16), along with the field equations, describe the interaction of a system of particles having isotopic spins.

An immediate consequence of Eq. (7.2.16) is

$$\frac{d}{d\tau}\mathbf{I}^2 = 0.$$

(7.2.18)

Hence the isotopic spin of each particle performs a precessional motion.

7.3 *The equations of motion according to Drechsler and Rosenblum*

In this section we give another derivation to the equations of motion of a non-Abelian charged particle from the principle of conservation of energy, momentum, and the isotopic current [113]. The equations of motion obtained are identical in structure to those derived in the last section from the semi-classical limit of the Dirac-Yang-Mills theory.

Define the isovector source current for N classical non-Abelian charges by the expression [113]

$$\mathbf{j}_v = \sum_{n=1}^{N} \int_{-\infty}^{\infty} \mathbf{v}_{(n)v} \delta^{(4)}[x - \xi_{(n)}(\tau)] \, d\tau, \tag{7.3.1}$$

where $\xi_{(n)}^{\mu}(\tau)$ is the trajectory of the nth particle, as a function of the proper time τ, and $\mathbf{v}_{(n)\mu}(\tau)$ is an isovector, space-time four-vector for the nth particle which will be shown to have the form

$$\mathbf{v}_{(n)}^{\mu}(\tau) = \mathbf{q}_{(n)}(\tau)\dot{\xi}_{(n)}^{\mu}(\tau), \tag{7.3.2}$$

with $\mathbf{q}_{(n)}(\tau)$ being the SU(2) charge at the proper time τ. As a consequence of the field equations, the current in Eq. (7.3.1) is covariantly conserved, that is

$$\nabla^v \mathbf{j}_v = \partial^v \mathbf{j}_v + \mathbf{b}^v \times \mathbf{j}_v = 0. \tag{7.3.3}$$

(Our potential has an opposite sign to that of Drechsler and Rosenblum [113].)

Besides the source current $\mathbf{j}_v$, one defines the symmetrical energy-momentum tensor of matter of N non-Abelian charged particles of masses $m_{(n)}$ in the following form:

$$t^{\mu v} = \sum_{n=1}^{N} \int_{-\infty}^{\infty} p_{(n)}^{\mu v}(\tau) \delta^{(4)}[x - \xi_{(n)}(\tau)] \, d\tau. \tag{7.3.4}$$

$p_{(n)}^{\mu v}(\tau)$ will be determined as a consequence of the energy and momentum conservation. In addition, the energy-momentum of the gauge field $\mathbf{f}_{\mu v}$ is given by

$$T_{(f)}^{\mu v} = \frac{1}{4}\eta^{\mu v}\mathbf{f}^{\kappa\lambda}\cdot\mathbf{f}_{\kappa\lambda} - \mathbf{f}^{\mu\kappa}\cdot\mathbf{f}_{\kappa}^{v}. \tag{7.3.5}$$

It is easy to show from the field equations that

$$\partial_\mu T_{(f)}^{\mu v} = -\mathbf{f}^{v\rho}\cdot\mathbf{j}_\rho. \tag{7.3.6}$$

The total energy and momentum conservation for a system of material particles and gauge fields is expressed as

$$\partial_\mu(t^{\mu\nu} + T^{\mu\nu}_{(f)}) = 0. \tag{7.3.7}$$

Using Eq. (7.3.6), the last equation will be

$$\partial_\mu t^{\mu\nu} - \mathbf{f}^{\nu\rho} \cdot \mathbf{j}_\rho = 0. \tag{7.3.8}$$

To derive the equations of motion of the particles, one uses Eqs. (7.3.1) and (7.3.4) in Eqs. (7.3.3) and (7.3.8), respectively, multiplying the first equation by an arbitrary smooth function $S(x)$ and the second by a set of arbitrary smooth functions $S_\nu(x)$, and integrating over all space-time. One obtains

$$\sum_{n=1}^{N} \iint_{-\infty}^{\infty} \left(\mathbf{v}^\nu_{(n)} \partial_\nu \delta^{(4)}[x - \xi_{(n)}(\tau)] \right.$$

$$\left. + \mathbf{b}_\nu \times \mathbf{v}^\nu_{(n)} \delta^{(4)}[x - \xi_{(n)}(\tau)] \right) S(x) \, d\tau \, d^4x = 0, \tag{7.3.9}$$

$$\sum_{n=1}^{N} \iint_{-\infty}^{\infty} \left(p^{\mu\nu}_{(n)} \partial_\mu \delta^{(4)}[x - \xi_{(n)}(\tau)] \right.$$

$$\left. - \mathbf{f}^{\nu\rho} \cdot \mathbf{v}_{(n)\rho} \delta^{(4)}[x - \xi_{(n)}(\tau)] \right) S_\nu(x) \, d\tau \, d^4x = 0. \tag{7.3.10}$$

Demanding now that the functions $S(x)$ and $S_\nu(x)$ vanish at the limits of the x and τ integrations, and performing an integration by parts on the first terms on the left-hand sides of Eqs. (7.3.9) and (7.3.10) over d^4x, one obtains

$$\sum_{n=1}^{N} \int_{-\infty}^{\infty} [\mathbf{v}^\nu_{(n)} \partial_\nu S - \mathbf{b}_\nu \times \mathbf{v}^\nu_{(n)} S]_{x=\xi_{(n)}(\tau)} \, d\tau = 0, \tag{7.3.11}$$

$$\sum_{n=1}^{N} \int_{-\infty}^{\infty} [p^{\mu\nu}_{(n)} \partial_\mu S_\nu + \mathbf{f}^{\nu\rho} \cdot \mathbf{v}_{(n)\rho} S_\nu]_{x=\xi_{(n)}(\tau)} \, d\tau = 0. \tag{7.3.12}$$

Following Mathisson [125], one breaks $p_{\mu\nu}$ and $\mathbf{v}^\mu_{(n)}$ into components parallel and perpendicular to the world line at a point determined by τ. To this end one defines, in addition to the velocity vectors $\dot{\xi}^\mu_{(n)}(\tau) = \dot{\xi}^\mu_{(n)}$, the space-like unit vectors $n^\mu_{(n)}(\tau) = n^\mu_{(n)}$ satisfying for each n,

$$n^\mu_{(n)} n_{(n)\mu} = -1, \qquad \dot{\xi}^\mu_{(n)} n_{(n)\mu} = 0, \tag{7.3.13}$$

and hence

$$v^\nu_{(n)} = \mathbf{q}_{(n)} \dot{\xi}^\nu_{(n)} + \mathbf{I}^\nu_{(n)}, \tag{7.3.14}$$

$$p^{\mu\nu}_{(n)} = m_{(n)} \dot{\xi}^\mu_{(n)} \dot{\xi}^\nu_{(n)} + a(n^\mu_{(n)} \dot{\xi}^\nu_{(n)} + n^\nu_{(n)} \dot{\xi}^\mu_{(n)}) + P^{\mu\nu}_{(n)}, \tag{7.3.15}$$

with the expressions $\mathbf{I}^\nu_{(n)}$ and $P^{\mu\nu}_{(n)} = P^{\nu\mu}_{(n)}$ satisfying

$$\mathbf{I}^\nu_{(n)} \dot{\xi}_{(n)\nu} = 0, \tag{7.3.16}$$

$$P^{\mu\nu}_{(n)} \dot{\xi}_{(n)\nu} = 0. \tag{7.3.17}$$

Using Eqs. (7.3.14) and (7.3.15) in Eqs. (7.3.11) and (7.3.12), one then gets

$$\sum_{n=1}^{N} \int_{-\infty}^{\infty} \{ [\mathbf{q}_{(n)} \dot{\xi}^\nu_{(n)} + \mathbf{I}^\nu_{(n)}] \partial_\nu S$$

$$- \mathbf{b}_\nu \times [\mathbf{q}_{(n)} \dot{\xi}^\nu_{(n)} + \mathbf{I}^\nu_{(n)}] S \}_{x=\xi_{(n)}(\tau)} d\tau = 0, \tag{7.3.18}$$

$$\sum_{n=1}^{N} \int_{-\infty}^{\infty} \{ [m_{(n)} \dot{\xi}^\mu_{(n)} \dot{\xi}^\nu_{(n)} + a(n^\mu_{(n)} \dot{\xi}^\nu_{(n)} + n^\nu_{(n)} \dot{\xi}^\mu_{(n)}) + P^{\mu\nu}_{(n)}] \partial_\mu S_\nu$$

$$+ \mathbf{f}^{\nu\rho} \cdot [\mathbf{q}_{(n)} \dot{\xi}_{(n)\rho} + \mathbf{I}_{(n)\rho}] S_\nu \}_{x=\xi_{(n)}(\tau)} d\tau = 0. \tag{7.3.19}$$

Since $\dot{\xi}^\mu_{(n)} \partial_\mu$ for $x = \xi_{(n)}(\tau)$ is equal to $d/d\tau$ taken along the nth world line, one can perform an integration by parts to obtain

$$\sum_{n=1}^{N} \int_{-\infty}^{\infty} \{ [(d/d\tau) \mathbf{q}_{(n)} + \mathbf{b}_\nu \times \mathbf{q}_{(n)} \dot{\xi}^\nu_{(n)}] S$$

$$- [\mathbf{I}^\nu_{(n)} \partial_\nu S - \mathbf{b}_\nu \times \mathbf{I}^\nu_{(n)} S] \}_{x=\xi_{(n)}(\tau)} d\tau = 0, \tag{7.3.20}$$

$$\sum_{n=1}^{N} \int_{-\infty}^{\infty} \left\{ \left[\frac{d}{d\tau} (m_{(n)} \dot{\xi}_{(n)}^{\nu} + a n_{(n)}^{\nu}) \right] S_{\nu} - [a n_{(n)}^{\mu} \dot{\xi}_{(n)}^{\nu} + P_{(n)}^{\mu\nu}] \partial_{\mu} S_{\nu} \right.$$

$$\left. - \mathbf{f}^{\nu\rho} \cdot [\mathbf{q}_{(n)} \dot{\xi}_{(n)\rho} + \mathbf{I}_{(n)\rho}] S_{\nu} \right\}_{x = \xi_{(n)}(\tau)} d\tau = 0. \tag{7.3.21}$$

Due to the arbitrariness of S_{ν}, $\partial_{\mu} S_{\nu}$, S and $\partial_{\mu} S$ at the point $x = \xi_{(n)}(\tau)$, one gets

$$\mathbf{I}_{(n)\nu} = 0, \tag{7.3.22}$$

and

$$a n_{(n)}^{\mu} \dot{\xi}_{(n)}^{\nu} + P_{(n)}^{\mu\nu} = 0. \tag{7.3.23}$$

Multiplying the last equation by $\dot{\xi}_{(n)\nu}$ gives $a = 0$. Then, from the same equation (7.3.23) one obtains $P_{(n)}^{\mu\nu} = 0$, and the final equations are

$$(d/d\tau) \mathbf{q}_{(n)} + \mathbf{b}_{\nu} \times \mathbf{q}_{(n)} \dot{\xi}_{(n)}^{\nu} = 0, \tag{7.3.24}$$

$$(d/d\tau)(m_{(n)} \dot{\xi}_{(n)}^{\nu}) = \mathbf{q}_{(n)} \cdot \mathbf{f}^{\nu\rho} \dot{\xi}_{(n)\rho}. \tag{7.3.25}$$

These are the Drechsler and Rosenblum equations of motion for non-Abelian charge. Equation (7.3.24) describes the covariant gauge charge precession, and Eq. (7.3.25) represents the Lorentz-type equation for the trajectory of the motion. For a single charge, dropping the subscript (n), one obtains

$$\frac{d\mathbf{q}}{d\tau} + \dot{\xi}^{\nu} \mathbf{b}_{\nu} \times \mathbf{q} = 0, \tag{7.3.26}$$

$$\frac{d(m \dot{\xi}^{\mu})}{d\tau} = \mathbf{q} \cdot \mathbf{f}^{\mu\rho} \dot{\xi}_{\rho}. \tag{7.3.27}$$

Equations (7.3.26) and (7.3.27) are identical to those of Wong (7.2.15) and (7.2.16) if we take $\mathbf{I} = \mathbf{q}$ and g is reinstated.

7.4 *The motion of a particle in the* SU(2) *magnetic field*

In the last two sections, we gave the equations of motion of a test particle, moving in the classical Yang-Mills field, in both the internal and the Minkowskian spaces. Now we present solutions of these equations when the Yang-Mills potentials and fields are those of the SU(2) magnetic field given by Eqs. (6.2.39) and (6.2.40) (see Ref. 91). A straightforward calculation gives

$$m\ddot{\xi}^0 = 0, \tag{7.4.1}$$

$$m\ddot{\xi}^1 = 0, \tag{7.4.2}$$

$$m\ddot{\xi}^2 = -\alpha \cdot \mathbf{I}\dot{\xi}^3, \tag{7.4.3}$$

$$m\ddot{\xi}^3 = \alpha \cdot \mathbf{I}\dot{\xi}^2, \tag{7.4.4}$$

and

$$\dot{I}_k + ge_{ijk}b_i I_j = 0, \tag{7.4.5}$$

along with $j^\mu = 0$. In Eq. (7.4.5) $b_i = b_i^\mu \dot{\xi}_\mu$, and α is given by Eq. (6.2.19).
Multiplying Eq. (7.4.5) by I_k one finds that $\mathbf{I} \cdot \dot{\mathbf{I}} = 0$, and therefore

$$d\mathbf{I}^2/d\tau = 0. \tag{7.4.6}$$

Consequently

$$\mathbf{I} = R\mathbf{n}, \tag{7.4.7}$$

where R is independent of τ, and $\mathbf{n}$ is the unit vector given by

$$\mathbf{n} = (\sin\theta\cos\phi, \sin\theta\sin\phi, \cos\theta). \tag{7.4.8}$$

Thus the particle moves in a sphere with radius R in the isospin space, and one has

$$\alpha \cdot \mathbf{I} = KR\sin\theta\cos(\omega - \phi). \tag{7.4.9}$$

To investigate the motion of the particle in the ordinary space-time, one has to solve Eqs. (7.4.1)–(7.4.4). The first two equations are trivial and their solutions are given by

$$\xi^0 = C_1 \tau + C_2, \tag{7.4.10}$$

$$\xi^1 = C_3 \tau + C_4, \tag{7.4.11}$$

where C_i are arbitrary constants. Differentiating Eqs. (7.4.3) and (7.4.4) with respect to τ, and using the notation

$$\dot{\xi}^2 = \eta, \qquad \dot{\xi}^3 = \zeta, \tag{7.4.12}$$

then one gets

$$\ddot{\eta} - \frac{\dot{f}}{f}\dot{\eta} + f^2\eta = 0, \tag{7.4.13}$$

$$\ddot{\zeta} - \frac{\dot{f}}{f}\dot{\zeta} + f^2\zeta = 0, \tag{7.4.14}$$

where $f = (1/m)\boldsymbol{\alpha} \cdot \mathbf{I}$ is a function of the space-time coordinates x.
    Introducing now the new parameter

$$s = \int f \, d\tau, \tag{7.4.15}$$

then Eqs. (7.4.13) and (7.4.14) are reduced to the simplified forms

$$\frac{d^2\eta}{ds^2} + \eta = 0, \tag{7.4.16}$$

$$\frac{d^2\zeta}{ds^2} + \zeta = 0. \tag{7.4.17}$$

The solutions of these equations are then given by

$$\eta = E_1 \cos s + E_2 \sin s, \tag{7.4.18}$$

$$\zeta = -E_2 \cos s + E_1 \sin s, \tag{7.4.19}$$

where E_1 and E_2 are arbitrary constants. One sees that the velocities of the particle, $\dot\xi^2$ and $\dot\xi^3$, in these two coordinates, behave like the coordinates of harmonic oscillators when the parameter s is used.

To find ξ^2 and ξ^3, one has to integrate Eqs. (7.4.18) and (7.4.19) with respect to τ. Then one obtains, using $d\tau = \dot s^{-1} ds$,

$$\xi^2 = \int \frac{\eta(s)}{\dot s} ds + \xi_0^2, \tag{7.4.20}$$

$$\xi^3 = \int \frac{\zeta(s)}{\dot s} ds + \xi_0^3, \tag{7.4.21}$$

where ξ_0^2 and ξ_0^3 are constants.

The integrals in the last two equations are functions of the parameter s which, by turn, depends on the function f. Now we consider particular cases of f.

(1) Let us first take

$$f = \Omega, \tag{7.4.22}$$

where Ω is a constant, and therefore the angle between the two isospin vectors **I** and $\boldsymbol{\alpha}$ is constant. Then one obtains

$$\eta = E_1 \cos \Omega\tau + E_2 \sin \Omega\tau, \tag{7.4.23}$$

$$\zeta = -E_2 \cos \Omega\tau + E_2 \sin \Omega\tau. \tag{7.4.24}$$

Hence ξ^2 and ξ^3 are given by

$$\xi^2 = \frac{1}{\Omega}(E_1 \sin \Omega\tau - E_2 \cos \Omega\tau) + \xi_0^2, \tag{7.4.25}$$

$$\xi^3 = \frac{-1}{\Omega}(E_2 \sin \Omega\tau + E_1 \cos \Omega\tau) + \xi_0^3. \tag{7.4.26}$$

In this case, one sees that the particle performs a harmonic motion with a constant frequency Ω.

(2) We now take

$$\frac{\dot{f}}{f} = \kappa, \tag{7.4.27}$$

where κ is a constant. One then obtains

$$f = f_0 e^{\kappa \tau}, \tag{7.4.28}$$

and

$$s = \kappa^{-1} f, \tag{7.4.29}$$

where f_0 is a constant. Using these results in Eqs. (7.4.20) and (7.4.21), and performing an integration of these equations, one gets

$$\xi^2 = \frac{E_1}{\kappa}\left[\ln s + \sum_{n=1}^{\infty} \frac{(-1)^n s^{2n}}{2n(2n)!}\right] + \frac{E_2}{\kappa}\sum_{n=0}^{\infty} \frac{(-1)^n s^{2n+1}}{(2n+1)(2n+1)!}, \tag{7.4.30}$$

$$\xi^3 = -\frac{E_2}{\kappa}\left[\ln s + \sum_{n=1}^{\infty} \frac{(-1)^n s^{2n}}{2n(2n)!}\right] + \frac{E_1}{\kappa}\sum_{n=0}^{\infty} \frac{(-1)^n s^{2n+1}}{(2n+1)(2n+1)!}. \tag{7.4.31}$$

Equations (7.4.30) and (7.4.31) can, of course, be written by means of the parameter τ.

7.5 *The motion of a particle in the instanton field*

In this section, we investigate the motion of a classical particle in the instanton field [92], which is a solution of the Yang-Mills field equations in the Euclidean space E^4. This solution, found by Belavin *et al.* [67], is non-singular and localized symmetrically in all directions in E^4. It is self-dual, that is, it carries zero energy, and it is characterized by a topological charge $q = 1$ (determined by the mapping $S^3 \to S^3$ covering the latter sphere one time). Furthermore, exact solutions describing an arbitrary number of instantons have been found with topological charges equal to the number of the instantons [68]. Thus, the instantons are considered as objects that exist in the Yang-Mills theory and could lead to physical results. Since these objects carry zero energy, it seems that they are some kind of vacuum fluctuations.

The instanton solution is given by [75]

$$b_0^j = \mp 2x^j/g(x^2 + V^2),\tag{7.5.1}$$

$$b_i^j = [2/g(x^2 + V^2)](-\varepsilon_{ijk}x^k \pm \delta_{ji}x^0),\tag{7.5.2}$$

$$f_{0i}^j = \mp 4\delta_{ji}V^2/g(x^2 + V^2)^2,\tag{7.5.3}$$

$$f_{ki}^j = 4\varepsilon_{jki}V^2/g(x^2 + V^2)^2,\tag{7.5.4}$$

$$x^2 = x_0^2 + x_1^2 + x_2^2 + x_3^2.\tag{7.5.5}$$

Here V is the size of the instanton, and the lower (upper) sign in Eqs. (7.5.1)–(7.5.4) gives the instanton (anti-instanton) solution with a topological charge $q = +1$ ($q = -1$).

To write the equations of motion in the Euclidean space one performs the following coordinate transformations $x \to x'$:

$$x'^0 = x^0, \qquad x'^k = -ix^k,\tag{7.5.6}$$

$$x_0' = x_0, \qquad x_k' = ix_k,\tag{7.5.7}$$

where x are the Minkowskian coordinates satisfying $x^\mu x_\mu = x_0^2 - x^2$, $\mathbf{x} = (x_1, x_2, x_3)$, $\mathbf{x}$ is a vector in the three-dimensional space, and $x^0 = x_0$, $x^k = -x_k$. Thus, $x'^\mu = x_\mu'$, $x'^\mu x_\mu' = x_0'^2 + x_1'^2 + x_2'^2 + x_3'^2$, where x' are the Euclidean coordinates.

Consequently, the equations of motion of the particle in the instanton field in the Euclidean space are given by (primes are omitted for brevity):

$$m\ddot{x}_0 = \mp [4V^2/(x^2 + V^2)^2]\mathbf{I} \cdot \dot{\mathbf{x}},\tag{7.5.8}$$

$$m\ddot{\mathbf{x}} = [4V^2/(x^2 + V^2)^2](\dot{\mathbf{x}} \times \mathbf{I} \pm \dot{x}_0\mathbf{I}),\tag{7.5.9}$$

$$\mathbf{I} = -[2/(x^2 + V^2)][\pm x_0(\dot{\mathbf{x}} \times \mathbf{I}) \mp \dot{x}_0(\mathbf{x} \times \mathbf{I}) + (\mathbf{I} \cdot \dot{\mathbf{x}})\mathbf{x} - (\mathbf{I} \cdot \mathbf{x})\dot{\mathbf{x}}].\tag{7.5.10}$$

Taking the vector product of x and Eq. (7.5.9), one gets, for $\mathbf{M} = m(\mathbf{x} \times \dot{\mathbf{x}})$, where $\mathbf{M}$ is the angular momentum of the particle, the following equation:

$$\dot{\mathbf{M}} = [4V^2/(x^2 + V^2)^2][(\mathbf{x} \cdot \mathbf{I})\dot{\mathbf{x}} - (\mathbf{x} \cdot \dot{\mathbf{x}})\mathbf{I} \pm \dot{x}_0(\mathbf{x} \times \mathbf{I})]. \qquad (7.5.11)$$

To understand the physical meaning of the above equations, one considers a particular case of $\mathbf{I}$ for which $I_1 = I_2 = 0$ and $I_3 \neq 0$. In this case, Eq. (7.5.10) leads to

$$I_3 = \text{constant}, \qquad (7.5.12)$$

and

$$x_1^2 + x_2^2 = k^2(x_0^2 + x_3^2), \qquad (7.5.13)$$

where k is a constant. From Eq. (7.5.11), one gets

$$\dot{M}_1 = [4V^2 I_3/(x^2 + V^2)^2](x_3\dot{x}_1 \pm x_2\dot{x}_0), \qquad (7.5.14)$$

$$\dot{M}_2 = [4V^2 I_3/(x^2 + V^2)^2](x_3\dot{x}_2 \mp x_1\dot{x}_0), \qquad (7.5.15)$$

$$\dot{M}_3 = -[2V^2 I_3/(x^2 + V^2)^2](d/d\tau)(x_1^2 + x_2^2). \qquad (7.5.16)$$

Equation (7.5.13) describes a hypersurface in which the particle moves. In the special case when $x_1^2 + x_2^2 = \text{constant}$, $(d/d\tau)(x_1^2 + x_2^2) = 0$, and thus one obtains

$$\dot{M}_3 = 0, \qquad (7.5.17)$$

by Eq. (7.5.16). Consequently, the angular momentum along the z direction is conserved.

If M_3 and M_1 are constants (M_2 is constant) too, then M_2 is constant (M_1 is constant), and

$$\cdot \ \dot{x}_1/x_2 = \mp\dot{x}_0/x_3. \qquad (7.5.18)$$

It is easy to see that the cases $I_2 = I_3 = 0$, $I_1 \neq 0$ and $I_3 = I_1 = 0$, $I_2 \neq 0$, give the same results, with the appropriate change of the axes.

In the more general case, one can take $I_1, I_2 \neq 0$ and $I_3 = 0$, for instance. Then the equations of motion become much more complicated, and are given by

$$\dot{I}_1 = -[2I_2/(x^2 + V^2)][\pm(\dot{x}_0 x_3 - \dot{x}_3 x_0) + \dot{x}_2 x_1 - x_2 \dot{x}_1], \qquad (7.5.19)$$

$$\dot{I}_2 = -(I_1/I_2)\dot{I}_1, \qquad (7.5.20)$$

$$\frac{I_1}{I_2} = \frac{\pm(\dot{x}_1 x_0 - \dot{x}_0 x_1) + \dot{x}_2 x_3 - x_2 \dot{x}_3}{\pm(\dot{x}_2 x_0 - \dot{x}_0 x_2) + \dot{x}_3 x_1 - x_3 \dot{x}_1}, \qquad (7.5.21)$$

$$\dot{M}_1 = [4V^2 I_2/(x^2 + V^2)^2]$$

$$\times [(-I_1/2I_2)d/d\tau(x_2^2 + x_3^2) + \dot{x}_2 \dot{x}_1 \mp x_3 \dot{x}_0], \qquad (7.5.22)$$

$$\dot{M}_2 = [4V^2 I_2/(x^2 + V^2)^2]$$

$$\times [-\tfrac{1}{2}(d/d\tau)(x_1^2 + x_3^2) + (I_1/I_2)(x_1 \dot{x}_2 \pm x_3 \dot{x}_0)], \qquad (7.5.23)$$

$$\dot{M}_3 = [4V^2 I_2/(x^2 + V^2)^2][(I_1/I_2)(x_1 \dot{x}_3 \mp x_2 \dot{x}_0) + x_2 \dot{x}_3 \pm x_1 \dot{x}_0]. \qquad (7.5.24)$$

No simple solutions were found to Eqs. (7.5.19)–(7.5.24) and further investigations should be done.

7.6 Summary

In this chapter, we presented the motion of a particle in classical Yang-Mills fields, which has recently attracted a good ideal of attention. We considered two cases of this problem. In the first case the particle moves in the field of the magnetic solution given in Chapter 6, whereas in the second the particle moves in the instanton field. In the two cases, one can see that the obtained equations of motion are complicated and hence only particular cases of them were solved.

For the case of the motion in the magnetic solution one can conclude that:

(1) The motion in the coordinates ξ^2 and ξ^3 of the ordinary space-time is harmonic when the product $\boldsymbol{\alpha} \cdot \mathbf{I}$ is constant. (This means that the angle between the direction of the field in the isospin space and the isovector $\mathbf{I}$, which

describes the motion of the particle in this space, is constant.) The motion in this case is analogous to that of an electric charge moving in a constant and uniform ordinary magnetic field directed along the x^1 axis [126].

(2) When $\alpha \cdot I$ is not constant, the motion in the mentioned coordinates behaves like the motion of damped harmonic oscillators, with the damping coefficient and the frequency not constant.

In the second case, the equations of motion in the instanton field were written by means of the angular momentum of the particle. These equations have been solved when two components of the isovector I are taken to be zero. The hypersurface in which the particle moves is described by the following relation between the coordinates:

$$x_1^2 + x_2^2 = k^2(x_0^2 + x_3^2).$$

In general, the equations of motion of the particle in space-time depend only on the field strengths and not on the potentials, whereas the equations of motion in the isospin space depend only on the potential. In the isospin space the particle performs a precessional motion for each value of the potentials.

8
Summary and Concluding Remarks

In this monograph we reviewed some exact solutions of the Yang-Mills field equations and their classification. Also, we discussed the motion of a test particle in classical Yang-Mills fields.

Exact solutions have an important role in quantum theory, and many papers were published in this area, giving new solutions. Most of these solutions were obtained for special cases only, and the field equations were not solved in the general case. The Yang-Mills field equations are highly nonlinear, and naturally one looks for some methods that simplify the calculations. One of the known methods is the classification of the Yang-Mills fields (see Chapter 3). Each type of field has some symmetry which leads to simplified equations. The use of the classification in general relativity theory enabled one to obtain all type D solutions. Also, in gauge theory, all class G solutions were obtained using classification methods.

In Chapter 3, we reviewed three known classification methods. The first is that of Carmeli in which the fields are classified according to the distinct eigenspinors and eigenvalues describing the Yang-Mills field, utilizing the gauge invariants of the field. This method is gauge invariant (see Sec. 3.1). The second is of Wang and Yang, in which the fields are classified according to the rank of the matrix describing the field. This method is Lorentz invariant (see Sec. 3.2). A comparison of these two methods was also given. The third is of Castillejo, Kugler and Roskies, in which the fields are classified according to the little group of gauge and Lorentz leaving the Yang-Mills field invariant (see Sec. 3.4).

Another method that simplifies the field equations is their formulation by means of the null tetrads. This was done by Carmeli, Charach and Kaye, and also by Newman (see Sec. 2.2). In this formalism, the symmetry of the equations appears and yields some simplification. This method enabled Carmeli and others to obtain exact solutions of the gauge field equations (see Chapters 4 and 5).

Using both the classification and the null-tetrad formalism, other exact solutions were obtained (see Chapter 6).

For a better understanding of the physical meaning of the exact solutions, one examines the motion of a test particle in these fields. Recently, several works were published on this topic and the equations of motion of the particle in classical Yang-Mills fields were derived. These equations were solved here for some Yang-Mills fields (see Chapter 7).

Monopole solutions of the sourceless Yang-Mills field equations and their comparison were also presented. The first is Carmeli's, obtained using the null-tetrad methods and assuming the potentials to have r^{-1} dependence. This solution describes a monopole that has both electric and magnetic charges. The second is Morris's, obtained using the Kalb ansatz and the assumption of time independence to all the field variables. This solution describes a magnetic monopole and is a particular case of the Georgi-Glashow model in the limit where the Higgs field vanishes. It was shown that, under certain conditions, these two solutions coincide, although the methods and the assumptions leading to them are different (see Chapter 4).

As an example for the application of the classification techniques, we classified in Chapter 5, some gauge fields which are solutions of one of Yang-Mills field equations and the Bianchi identity. These solutions, obtained by Altamirano and Villarroel, were classified using the spinorial method (see Sec. 3.1). The classification is done according to the distinct number of the eigenspinors and eigenvalues describing the gauge fields, and according to the relations between the gauge-group invariants (see Chapter 5).

The other solutions, presented in Chapter 6, are obtained using the classification of the fields and the null-tetrad formalism. The Yang-Mills field equations are solved for two classes of the Castillejo-Kugler-Roskies classification (see Sec. 3.4). The first is class A (see Table 3.4.1). Substituting the general form of the fields belonging to this class, in the field equations written by means of the null tetrads, and analyzing these equations, yields the magnetic solution (the electric part vanishes). It is given by

$$f_{\mu\nu}^k = \frac{1}{g} f_{\mu\nu} \alpha^k,$$

where $f_{\mu\nu}$ are constants and α^k is a vector in the isospin space with a constant magnitude. Although the field strengths are Abelian, the potentials are not (their explicit form is given in Sec. 6.2).

This solution was classified according to the two classification methods and was found to be of type D_p of Carmeli (see Fig. 3.1.2), and to subcase 3_b of Wang-Yang (see Sec. 3.2), classifications. The energy density of this solution was calculated, and it was found to be a constant throughout the space-time. Also the motion of a particle in this field was investigated.

The other exact solutions considered in this review belong to class C_1 of the Castillejo-Kugler-Roskies classification (see Table 3.4.1). The matrix $A = E + iH$, describing these fields, is diagonal and two of its elements are equal. Two sets of solutions of class C_1 and their corresponding energy density were presented (see Sec. 6.4).

We also reviewed two independent methods deriving the equations of motion of a particle. The first is of Wong, where the equations of motion are obtained taking an appropriate limit to the quantum fields. The second is of Drechsler and Rosenblum who obtained the same equations of motion from energy and momentum, and covariant isotopic charge, conservation. One of these equations is a generalization to the Lorentz equation for a charged particle moving in an electromagnetic field in space-time. The other describes a precessional motion in the isospin space (see Secs. 7.2 and 7.3).

The motion of a particle was discussed for two cases of Yang-Mills fields. The first is in the magnetic solution given in Chapter 6. It was shown that the motion, when the angle between the isovector and the Yang-Mills field is constant, is analogous to that of an electric charge moving in a constant and uniform ordinary magnetic field. The more general case was also discussed.

The motion in another field, the instanton, also examined, is of great interest from the physical point of view. The equations of motion of a particle moving in this field were given and expressed by means of the angular momentum of the particle. Particular exact solutions for the motion in this field, when two components of the isospin vector **I** vanish, were given. It was found that the particle moves in a hypersurface given by

$$x_1^2 + x_2^2 = k^2(x_0^2 + x_3^2),$$

where k is constant and x_μ are the coordinates. The equations of motion for the more general case, along with their angular momenta, were also presented (see Sec. 7.5).

In conclusion, the following three major topics were reviewed:

(1) Exact solutions of the Yang-Mills field equations using the null-tetrad method.

(2) The classification of these fields using three different methods.

(3) The motion of a particle in these fields, including the magnetic and the instanton fields.

Appendix A: Derivation of Equations (6.2.1)–(6.2.4)

In this Appendix, we give the derivation of Eqs. (6.2.1)–(6.2.4). Firstly, one has to calculate the fields χ_{ack} in polar coordinates, which are given by Eq. (2.2.19). From Eqs. (2.2.4) and (2.2.10) χ_{ack} have the form

$$\chi_{ack} = \tfrac{1}{2} \varepsilon^{b'd'} \sigma_{ab'}^{\mu} \sigma_{cd'}^{\nu} f_{\mu\nu k}. \tag{A.1}$$

Using Eqs. (2.2.3) and (2.2.20) in Eq. (A.1), one obtains

$$\chi_{00k} = \chi_{0k} = \frac{1}{\sqrt{2}r}(f_{12k} + \frac{i}{\sin\theta}f_{13k}), \tag{A.2}$$

$$\chi_{01k} = \chi_{10k} = \chi_{1k} = \frac{1}{2}\left(f_{10k} + \frac{i}{r^2\sin\theta}f_{23k}\right), \tag{A.3}$$

$$\chi_{11k} = \chi_{2k} = \frac{1}{2\sqrt{2}r}\left[2f_{20k} + f_{12k} + \frac{i}{\sin\theta}(f_{31k} + 2f_{03})\right]. \qquad \text{(A.4)}$$

The fields $f_{\mu\nu k}$ appearing in Eqs. (A.1)–(A.4) are in the polar coordinates given by Eq. (2.2.19). Thus, one has to transform the fields, which are given by Table 3.4.1, from Cartesian coordinates to polar ones.

From Table 3.4.1, the electric and magnetic parts of the Yang-Mills fields, for class A, are

$$E = \begin{pmatrix} a & 0 & 0 \\ 0 & 0 & 0 \\ 0 & 0 & 0 \end{pmatrix}, \qquad H = \begin{pmatrix} b & 0 & 0 \\ c & 0 & 0 \\ 0 & 0 & 0 \end{pmatrix}, \qquad \text{(A.5)}$$

where a, b and c are functions of the coordinates. Then the values of $f'_{\mu\nu k}$ are

$$f'_{10k} = E_{1k} = a\delta_1^k, \qquad \text{(A.6)}$$

$$f'_{23k} = H_{1k} = (b\delta_1^k + c\delta_2^k), \qquad \text{(A.7)}$$

all the other components vanish. In the last two equations $f'_{i0k} = E_{ik}$ and $f'_{ijk} = \varepsilon_{ijm}H_{mk}$.

One can transform the fields $f'_{\mu\nu k}$ to the polar coordinates according to the relation

$$f_{\mu\nu k} = \frac{\partial x'^{\alpha}}{\partial x^{\mu}}\frac{\partial x'^{\beta}}{\partial x^{\nu}}f'_{\alpha\beta k}, \qquad \text{(A.8)}$$

where $x'^{\alpha} = (t, x, y, z)$, $x^{\mu} = (u, r, \theta, \phi)$, $f'_{\mu\nu k}$ are in the Cartesian coordinate and $f_{\mu\nu k}$ are in the polar ones. The relations between the two coordinate systems are

$$t = u + r, \quad x = r\sin\theta\cos\phi, \quad y = r\sin\theta\sin\phi, \quad z = r\cos\theta. \qquad \text{(A.9)}$$

From Eqs. (A.8) and (A.9), one gets

$$f_{01k} = \sin\theta\cos\phi\, f'_{01k} + \sin\theta\sin\phi\, f'_{02k} + \cos\theta\, f'_{03k}, \qquad \text{(A.10)}$$

$$f_{02k} = r(\cos\theta\cos\phi\, f'_{01k} + \cos\theta\sin\phi\, f'_{02k} - \sin\theta\, f'_{03k}), \qquad \text{(A.11)}$$

$$f_{03k} = r \sin \theta (\cos \phi \, f'_{02k} - \sin \phi \, f'_{01k}), \tag{A.12}$$

$$f_{12k} = r(\cos \theta \cos \phi \, f'_{01k} + \cos \theta \sin \phi \, f'_{02k} - \sin \theta \, f'_{03k}$$

$$+ \cos \phi \, f'_{31k} + \sin \phi \, f'_{32k}), \tag{A.13}$$

$$f_{13k} = r \sin \theta (\cos \phi \, f'_{02k} - \sin \phi \, f'_{01k} + \sin \theta \, f'_{12k}$$

$$+ \cos \theta \sin \phi \, f'_{13k} + \cos \theta \cos \phi \, f'_{32k}), \tag{A.14}$$

$$f_{23k} = r^2 \sin \theta (\cos \theta \, f'_{12k} + \sin \theta \sin \phi \, f'_{31k} + \sin \theta \cos \phi \, f'_{23k}). \tag{A.15}$$

Using Eqs. (A.6) and (A.7) in the last equations one gets for class A

$$f_{01k} = -a\delta_1^k \sin \theta \cos \phi, \tag{A.16}$$

$$f_{02k} = -a\delta_1^k r \cos \theta \cos \phi, \tag{A.17}$$

$$f_{03k} = a\delta_1^k r \sin \theta \sin \phi, \tag{A.18}$$

$$f_{12k} = -a\delta_1^k r \cos \theta \cos \phi - (b\delta_1^k + c\delta_2^k) r \sin \phi, \tag{A.19}$$

$$f_{13k} = a\delta_1^k r \sin \theta \sin \phi - (b\delta_1^k + c\delta_2^k) r \sin \theta \cos \theta \cos \phi, \tag{A.20}$$

$$f_{23k} = (b\delta_1^k + c\delta_2^k) r^2 \sin^2 \theta \cos \phi. \tag{A.21}$$

Inserting the values of the fields $f_{\mu\nu k}$, for class A, from Eqs. (A.16)–(A.21) in Eqs. (A.2)–(A.4) one obtains for χ_{ack} the following:

$$\chi_{0k} = \frac{-1}{\sqrt{2}} (\cos \theta \cos \phi - i \sin \phi) \beta_k, \tag{A.22}$$

$$\chi_{1k} = \frac{1}{2} \sin \theta \cos \phi \, \beta_k, \tag{A.23}$$

$$\chi_{2k} = \frac{1}{2\sqrt{2}} (\cos \theta \cos \phi + i \sin \phi) \beta_k, \tag{A.24}$$

where β_k is an isospin vector given by

$$\beta_k = (a + ib, ic, 0). \tag{A.25}$$

Using the last four equations in Eqs. (2.2.27)–(2.2.30), one gets Eqs. (6.2.1)–(6.2.4). For example, to get Eq. (6.2.3) we put the values of χ_{0k}, χ_{1k} and χ_{2k} from Eqs. (A.22)–(A.24) in Eqs. (2.2.28) and (2.2.30). Equation (2.2.28) will take the form

$$(\mathcal{D}n_1)\left(\frac{\partial}{\partial u} - \frac{1}{2}\frac{\partial}{\partial r}\right)\beta_k = \frac{-n_1}{2r}\mathcal{D}\beta_k + g\varepsilon_{ijk}\left[(\mathcal{D}n_1)\beta_i b_{11'j} + \frac{n_1}{\sqrt{2}}\beta_i b_{01'j}\right], \tag{A.26}$$

while (2.2.30) will be

$$n_1\left(\frac{\partial}{\partial u} - \frac{1}{2}\frac{\partial}{\partial r}\right)\beta_k = \frac{(\bar{\mathcal{D}}n_1)}{2r}\mathcal{D}\beta_k + g\varepsilon_{ijk}\left[n_1\beta_i b_{11'j} - \frac{1}{\sqrt{2}}(\bar{\mathcal{D}}n_1)\beta_i b_{01'j}\right], \tag{A.27}$$

where n_k and $\mathcal{D}$ are given by Eqs. (6.4.7) and (2.2.26) respectively.

Multiplying Eq. (A.27) by $\mathcal{D}n_1$ and Eq. (A.26) by n_1 and subtracting the last equation from the first, one gets

$$0 = \frac{1}{2}[(\bar{\mathcal{D}}n_1)(\mathcal{D}n_1) + n_1^2]\left[\frac{\mathcal{D}\beta_k}{r} - \sqrt{2}g\varepsilon_{ijk}\beta_i b_{01'j}\right]. \tag{A.28}$$

But $(\bar{\mathcal{D}}n_1)(\mathcal{D}n_1) + n_1^2 = 1$, so, (A.28) is identical to Eq. (6.2.3). Inserting Eq. (A.28) into Eq. (A.26) or Eq. (A.27), one obtains Eq. (6.2.1).

In the same way one gets (6.2.4) and (6.2.2) from Eqs. (2.2.27) and (2.2.29). Using Eqs. (A.22)–(A.24), Eq. (2.2.27) becomes

$$n_1\frac{\partial}{\partial r}\beta_k = -\frac{(\mathcal{D}n_1)}{r}\bar{\mathcal{D}}\beta_k + g\varepsilon_{ijk}[\sqrt{2}(\mathcal{D}n_1)\beta_i b_{10'j} + n_1\beta_i b_{00'j}], \tag{A.29}$$

and Eq. (2.2.29) will be

$$(\bar{\mathscr{D}}n_1)\frac{\partial \beta_k}{\partial r} = \frac{n_1}{r}\bar{\mathscr{D}}\beta_k - g\varepsilon_{ijk}[\sqrt{2}n_1\beta_i b_{10'j} - (\bar{\mathscr{D}}n_1)\beta_i b_{00'j}]. \qquad (A.30)$$

Multiplying Eq. (A.29) by $\bar{\mathscr{D}}n_1$ and Eq. (A.30) by n_1 and subtracting the first equation from the last, one gets

$$0 = [n_1^2 + (\bar{\mathscr{D}}n_1)(\mathscr{D}n_1)]\left[\frac{\bar{\mathscr{D}}\beta_k}{r} - \sqrt{2}g\varepsilon_{ijk}\beta_i b_{10'j}\right]. \qquad (A.31)$$

So Eq. (A.31) is identical to Eq. (6.2.4). Inserting Eq. (A.31) into Eq. (A.29) or Eq. (A.30), one obtains Eq. (6.2.2).

Appendix B: Derivation of Equations (6.4.8)–(6.4.11)

Here we give the derivation of Eq. (6.4.8)–(6.4.11). We follow the same method used in Appendix A. Writing the general form of the fields χ_{ack} for Class C_1, then substituting the results in the Yang-Mills field equations.

From Table 3.4.1, the electric and magnetic parts of the Yang-Mills fields are given by

$$E_{1k} = a\delta_1^k, \qquad E_{2k} = b\delta_2^k, \qquad E_{3k} = b\delta_3^k,$$

$$H_{1k} = c\delta_1^k, \qquad H_{2k} = d\delta_2^k, \qquad H_{3k} = d\delta_3^k,$$

(B.1)

where a, b, c and d are arbitrary real functions of the coordinates.

Then the fields $f'_{\mu\nu k}$, in the Cartesian coordinates have the forms

$$f'_{10k} = a\delta_1^k, \qquad f'_{20k} = b\delta_2^k, \qquad f'_{30k} = b\delta_3^k,$$

$$f'_{12k} = d\delta_3^k, \qquad f'_{23k} = c\delta_1^k, \qquad f'_{31k} = d\delta_2^k. \tag{B.2}$$

The fields $f_{\mu\nu k}$ in the polar coordinates are

$$f_{10k} = (a \sin\theta \cos\phi, b \sin\theta \sin\phi, b \cos\theta) \tag{B.3}$$

$$f_{20k} = r(a \cos\theta \cos\phi, b \cos\theta \sin\phi, -b \sin\theta), \tag{B.4}$$

$$f_{30k} = r \sin\theta(-a \sin\phi, b \cos\phi, 0), \tag{B.5}$$

$$f_{12k} = r(-a \cos\theta \cos\phi - c \sin\phi, d \cos\phi - b \cos\theta \sin\phi, b \sin\theta), \tag{B.6}$$

$$f_{23k} = r^2 \sin\theta(c \sin\theta \cos\phi, d \sin\theta \sin\phi, d \cos\theta), \tag{B.7}$$

$$f_{31k} = r \sin\theta(-a \sin\phi + \cos\theta \cos\phi, b \cos\phi + d \cos\theta \sin\phi, -d \sin\theta). \tag{B.8}$$

Inserting the values of $f_{\mu\nu k}$ from the last equations in Eqs. (A.2)–(A.4), one obtains for the χ_{ack}

$$\chi_{0k} = \frac{-1}{\sqrt{2}} [(a + ic)\mathscr{D}\gamma_k + (b + id)\mathscr{D}\delta_k], \tag{B.9}$$

$$\chi_{1k} = \frac{1}{2} [(a + ic)\gamma_k + (b + id)\delta_k], \tag{B.10}$$

$$\chi_{2k} = \frac{1}{2\sqrt{2}} [(a + ic)\bar{\mathscr{D}}\gamma_k + (b + id)\bar{\mathscr{D}}\delta_k], \tag{B.11}$$

where γ_k and δ_k are given by Eqs. (6.4.5) and (6.4.6) respectively.

Substituting the fields χ_{ack} from Eqs. (B.9)–(B.11) in the Yang-Mills field equations (2.2.27)–(2.2.30) one obtains

$$\frac{\partial}{\partial r}[\gamma_k(a + ic) + \delta_k(b + id)]$$

$$= -\frac{1}{r}[(\mathcal{D}\gamma_k)\bar{\mathcal{D}}(a + ic) + (\mathcal{D}\delta_k)\bar{\mathcal{D}}(b + id)] + g\varepsilon_{ijk}\{\sqrt{2}[(a + ic)\mathcal{D}\gamma_i$$

$$+ (b + id)\mathcal{D}\delta_i]b_{10'j} + [(a + ic)\gamma_1 + (b + id)\delta_i]b_{00'j}\}, \tag{B.12}$$

$$\left(\frac{\partial}{\partial u} - \frac{1}{2}\frac{\partial}{\partial r}\right)[(\mathcal{D}\gamma_k)(a + ic) + (\mathcal{D}\delta_k)(b + id)]$$

$$= \frac{-1}{2r}[\gamma_k\mathcal{D}(a + ic) + \delta_k\mathcal{D}(b + id)] + g\varepsilon_{ijk}\{[(a + ic)\mathcal{D}\gamma_i$$

$$+ (b + id)\mathcal{D}\delta_i]b_{11'j} + \frac{1}{\sqrt{2}}[(a + ic)\gamma_i + (b + id)\delta_i]b_{01'j}\}, \tag{B.13}$$

$$\frac{\partial}{\partial r}[(a + ic)\bar{\mathcal{D}}\gamma_k + (b + id)\bar{\mathcal{D}}\delta_k]$$

$$= \frac{1}{r}[\gamma_k\bar{\mathcal{D}}(a + ic) + \delta_k\bar{\mathcal{D}}(b + id)] - g\varepsilon_{ijk}\{\sqrt{2}[(a + ic)\gamma_i$$

$$+ (b + id)\delta_i]b_{10'j} - [(a + ic)\bar{\mathcal{D}}\gamma_i + (b + id)\bar{\mathcal{D}}\delta_i]b_{00'j}\}, \tag{B.14}$$

$$\left(\frac{\partial}{\partial u} - \frac{1}{2}\frac{\partial}{2r}\right)[\gamma_k(a + ic) + \delta_k(b + id)]$$

$$= \frac{1}{2r}[(\bar{\mathcal{D}}\gamma_k)\mathcal{D}(a + ic) + \bar{\mathcal{D}}\delta_k)\mathcal{D}(b + id)] - g\varepsilon_{ijk}\left\{\frac{1}{\sqrt{2}}[(a + ic)\bar{\mathcal{D}}\gamma_i\right.$$

$$\left. + (b + id)\bar{\mathcal{D}}\delta_i]b_{01'j} - [(a + ic)\gamma_i + (b + id)\delta_i]b_{11'j}\right\}. \tag{B.15}$$

If we write the above equation in components, we get twelve equations. But we can also take the sum or the difference between the different components to get simplest forms. This is what we had done. For example, if one takes $k = 1$ in Eqs. (B.12) and (B.14), one will get respectively

$$n_1 \frac{\partial}{\partial r}(a + ic) = \frac{-\mathscr{D}n_1}{r} \bar{\mathscr{D}}(a + ic) + g(b + id)[\sqrt{2}(b_{10'3}\mathscr{D}n_2 - b_{10'2}\mathscr{D}n_3)$$

$$+ b_{00'3}n_2 - b_{00'2}], \tag{B.16}$$

$$(\bar{\mathscr{D}}n_1)\frac{\partial}{\partial r}(a + ic) = \frac{n_1}{r}\bar{\mathscr{D}}(a + ic) - g(b + id)[\sqrt{2}(b_{10'3}n_2 - b_{10'2}n_3)$$

$$- (b_{00'3}\bar{\mathscr{D}}n_2 - b_{00'2}\bar{\mathscr{D}}n_3)]. \tag{B.17}$$

[Remember that $\gamma_k = (n_1, 0, 0)$ and $\delta_k = (0, n_2, n_3)$].
Multiplying Eq. (B.16) by n_1 and Eq. (B.17) by $\mathscr{D}n_1$, then the sum of them yields

$$\frac{\partial}{\partial r}(a + ic) = -ig(b + id))[\sqrt{2}(b_{10'2}\mathscr{D}n_2 + b_{10'3}\mathscr{D}n_3) + b_{00'2}n_2 + b_{00'3}n_3].$$
$$\tag{B.18}$$

Also, multiplying Eq. (B.16) by $\bar{\mathscr{D}}n_1$ and Eq. (B.17) by n_1 and subtracting the last equation from the first, one obtains

$$\frac{1}{r}\bar{\mathscr{D}}(a + ic) = ig(b + id)[\sqrt{2}(b_{10'2}n_2 + b_{10'3}n_3) - (b_{00'2}\bar{\mathscr{D}}n_2 + b_{00'3}\bar{\mathscr{D}}n_3)].$$
$$\tag{B.19}$$

One can continue with this process for the other components of the Eqs. (B.12)–(B.15).
From Eqs. (B.12) and (B.14), when $k = 2$, one obtains

$$\frac{\partial}{\partial r}(b + id) = -ig[(a + ic)(\sqrt{2}b_{10'3}\mathscr{D}n_3 + b_{00'3}n_3)$$

$$+ (b + id)(\sqrt{2}b_{10'1}\mathscr{D}n_1 + b_{00'1}n_1)], \tag{B.20}$$

$$\frac{1}{r}\bar{\mathscr{D}}(b + id) = ig[(a + ic)(\sqrt{2}b_{10'3}n_3 - b_{00'3}\bar{\mathscr{D}}n_3)$$

$$+ (b + id)(\sqrt{2}b_{10'1}n_1 - b_{00'1}\bar{\mathscr{D}}n_1)]. \tag{B.21}$$

From the same two equations, when $k = 3$, one gets

$$\frac{\partial}{\partial r}(b + id) \stackrel{\pm}{=} -ig[(a + ic)(\sqrt{2}b_{10'2}\mathscr{D}n_2 + b_{00'2}n_2)$$

$$+ (b + id)(\sqrt{2}b_{10'1}\mathscr{D}n_1 + b_{00'1}n_1)], \tag{B.22}$$

$$\frac{1}{r}\mathscr{D}(b + id) = ig[(a + ic)(\sqrt{2}b_{10'2}n_2 - b_{00'2}\bar{\mathscr{D}}n_2)$$

$$+ (b + id)(\sqrt{2}b_{10'1}n_1 - b_{00'1}\bar{\mathscr{D}}n_1)]. \tag{B.23}$$

For the two other equations (B.13) and (B.16), when $k = 1$, the outcoming two equations are

$$\left(\frac{\partial}{\partial u} - \frac{1}{2}\frac{\partial}{\partial r}\right)(a + ic) = ig(b + id)[b_{11'2}n_2 + b_{11'3}n_3$$

$$- \frac{1}{\sqrt{2}}(b_{01'2}\bar{\mathscr{D}}n_2 + b_{01'3}\bar{\mathscr{D}}n_3)], \tag{B.24}$$

$$\frac{1}{2r}\mathscr{D}(a + ic) = -ig(b + id)\left[\frac{1}{\sqrt{2}}(b_{01'2}n_2 + b_{01'3}n_3)\right.$$

$$\left. + b_{11'2}\mathscr{D}n_2 + b_{11'3}\mathscr{D}n_3\right]. \tag{B.25}$$

For $k = 2$

$$\left(\frac{\partial}{\partial u} - \frac{1}{2}\frac{\partial}{\partial r}\right)(b + id) = ig\left[(a + ic)\left(b_{11'3}n_3 - \frac{1}{\sqrt{2}}b_{0'13}\bar{\mathscr{D}}n_3\right)\right.$$

$$\left. + (b + id)\left(b_{11'1}n_1 - \frac{1}{\sqrt{2}}b_{01'1}\bar{\mathscr{D}}n_1\right)\right], \tag{B.26}$$

$$\frac{1}{2r}\mathcal{D}(b+id) = -ig\left[(a+ic)\left(b_{11'3}\mathcal{D}n_3 + \frac{1}{\sqrt{2}}b_{01'3}n_3\right)\right.$$

$$\left. + (b+id)\left(b_{11'1}\mathcal{D}n_1 + \frac{1}{\sqrt{2}}b_{01'1}n_1\right)\right]. \quad \text{(B.27)}$$

For $k = 3$

$$\left(\frac{\partial}{\partial u} - \frac{1}{2}\frac{\partial}{\partial r}\right)(b+id) = ig\left[(a+ic)\left(b_{11'2}n_2 - \frac{1}{\sqrt{2}}b_{01'2}\bar{\mathcal{D}}n_2\right)\right.$$

$$\left. + (b+id)\left(b_{11'1}n_1 - \frac{1}{\sqrt{2}}b_{01'1}\bar{\mathcal{D}}n_1\right)\right], \quad \text{(B.28)}$$

$$\frac{1}{2r}\mathcal{D}(b+id) = -ig\left[(a+ic)\left(b_{11'2}\mathcal{D}n_2 + \frac{1}{\sqrt{2}}b_{01'2}n_2\right)\right.$$

$$\left. + (b+id)\left(b_{11'1}\mathcal{D}n_1 + \frac{1}{\sqrt{2}}b_{01'1}n_1\right)\right]. \quad \text{(B.29)}$$

Now, one can examine the last twelve equations (B.18)–(B.29) and see that Eqs. (6.4.8)–(6.4.11) are obtained directly from those equations. For example, the difference between Eqs. (B.20) and (B.22) is

$$(a+ic)\left[\sqrt{2}b_{10'3}\mathcal{D}n_3 + b_{00'3}n_3 - (\sqrt{2}b_{10'2}\mathcal{D}n_2 + b_{00'2}n_2)\right] = 0. \quad \text{(B.30)}$$

In the general case $a \neq 0$ and $c \neq 0$ then from the last equation (B.30) one gets Eq. (6.4.8). Also the difference between Eqs. (B.21) and (B.23) is

$$(a+ic)\left[\sqrt{2}b_{01'2}n_2 - b_{00'2}\bar{\mathcal{D}}n_2 - (\sqrt{2}b_{10'3}n_3 - b_{00'3}\bar{\mathcal{D}}n_3)\right] = 0. \quad \text{(B.31)}$$

So, from Eq. (B.31) one gets Eq. (6.4.9). In the same way, Eq. (6.4.10) can be obtained from Eqs. (B.26) and (B.28), also Eq. (6.4.11) is obtained from Eqs. (B.27) and (B.29).

Using Eqs. (6.4.8)–(6.4.11), Eqs. (B.18), (B.19), (B.24) and (B.25) will be as follows:

$$\frac{\partial}{\partial r}(a + ic) = -2ig(b + id)(\sqrt{2}b_{10'3}\mathscr{D}n_3 + b_{00'3}n_3), \tag{B.18a}$$

$$\frac{1}{r}\bar{\mathscr{D}}(a + ic) = 2ig(b + id)(\sqrt{2}b_{10'3}n_3 - b_{00'3}\bar{\mathscr{D}}n_3), \tag{B.19a}$$

$$\left(\frac{\partial}{\partial u} - \frac{1}{2}\frac{\partial}{\partial r}\right)(a + ic) = 2ig(b + id)\left(b_{11'3}n_3 - \frac{1}{\sqrt{2}}b_{01'3}\mathscr{D}n_3\right), \tag{B.24a}$$

$$\frac{1}{r}\mathscr{D}(a + ic) = -2ig(b + id)\left(b_{11'3}\mathscr{D}n_3 + \frac{1}{\sqrt{2}}b_{01'3}n_3\right). \tag{B.25a}$$

Appendix C: Derivation of Equations (6.4.20)–(6.4.26)

The algebraic equations (6.4.8)–(6.4.11) have two independent solutions Eqs. (6.4.12)–(6.4.19). The substituting of those solutions in the field equations and the relations between the fields and the potentials give the relations (6.4.20)–(6.4.26) for case 1. In this Appendix, we give more details about the process by which those relations are obtained.

We start by inserting the values of the potentials, for case 1 Eqs. (6.4.12)–(6.4.15), in the field equations (B.18)–(B.29), and obtain

$$\frac{\partial}{\partial r}(a + ic) = 2g(b + id)Fn_1,\tag{C.1}$$

$$\frac{1}{r}\bar{\mathscr{D}}(a + ic) = 2g(b + id)F\bar{\mathscr{D}}n_1,\tag{C.2}$$

$$\frac{1}{r}\mathcal{D}(a + ic) = 2g(b + id)F\mathcal{D}n_1, \tag{C.3}$$

$$\left(\frac{\partial}{\partial u} - \frac{1}{2}\frac{\partial}{\partial r}\right)(a + ic) = -g(b + id)Fn_1, \tag{C.4}$$

$$\frac{\partial}{\partial r}(b + id) = g(a + ic)Fn_1 - ig(b + id)(b_{00'1}n_1 + \sqrt{2}b_{10'1}\mathcal{D}n_1), \tag{C.5}$$

$$\frac{1}{r}\bar{\mathcal{D}}(b + id) = g(a + ic)F\bar{\mathcal{D}}n_1 + ig(b + id)(\sqrt{2}b_{10'1}n_1 - b_{00'1}\mathcal{D}n_1), \tag{C.6}$$

$$\frac{1}{r}\mathcal{D}(b + id) = g(a + ic)F\mathcal{D}n_1 - 2ig(b + id)\left(b_{11'1}\mathcal{D}n_1 + \frac{1}{\sqrt{2}}b_{01'1}n_1\right), \tag{C.7}$$

$$\left(\frac{\partial}{\partial u} - \frac{1}{2}\frac{\partial}{\partial r}\right)(b + id) = -\frac{1}{2}g(a + ic)Fn_1 + ig(b + id)\left(b_{11'1}n_1 - \frac{1}{\sqrt{2}}b_{01'1}\bar{\mathcal{D}}n_1\right), \tag{C.8}$$

where F is an arbitrary real function of the coordinates. If one adds one half of Eq. (C.1) to (C.4), one gets

$$\frac{\partial}{\partial u}(a + ic) = 0. \tag{C.9}$$

Thus $a + ic$ is independent of u.

Substituting the potential from Eqs. (6.4.12)–(6.4.15) in the relations between the fields and the potentials Eqs. (2.2.23)–(2.2.25), those relations in components become

$$\frac{-1}{\sqrt{2}}(a + ic + igF^2)\mathcal{D}n_1 = \frac{1}{\sqrt{2r}}\mathcal{D}b_{00'1} - \left(\frac{\partial}{\partial r} + \frac{1}{r}\right)b_{01'1}, \tag{C.10}$$

$$\frac{-1}{gF}\left[(b-id)n_1 - i\frac{\partial}{\partial r}F\right] = b_{00'1}n_1 + \sqrt{2}b_{10'1}\mathscr{D}n_1, \qquad \text{(C.11)}$$

$$\frac{-1}{gF}\left[(b-id)\mathscr{D}n_1 - \frac{i}{r}\bar{\mathscr{D}}F\right] = b_{00'1}\bar{\mathscr{D}}n_1 - \sqrt{2}b_{10'}n_1, \qquad \text{(C.12)}$$

$$\frac{1}{2\sqrt{2}}(a+ic+igF^2)\bar{\mathscr{D}}n_1 = \left(\frac{\partial}{\partial u} - \frac{1}{2}\frac{\partial}{\partial r} - \frac{1}{2r}\right)b_{10'1} - \frac{1}{\sqrt{2}r}\bar{\mathscr{D}}b_{11'1}, \qquad \text{(C.13)}$$

$$-\frac{1}{2gF}\left[(b-id)\mathscr{D}n_1 - \frac{i}{r}\mathscr{D}F\right] = b_{11'1}\mathscr{D}n_1 + \frac{1}{\sqrt{2}}b_{0'1}n_1 \qquad \text{(C.14)}$$

$$-\frac{1}{2gF}\left[(b-id)n_1 + 2i\left(\frac{\partial}{\partial u} - \frac{1}{2}\frac{\partial}{\partial r}\right)F\right] = b_{11'1}n_1 - \frac{1}{\sqrt{2}}b_{01'1}\bar{\mathscr{D}}n_1, \qquad \text{(C.15)}$$

$$(a+ic+igF^2)n_1 = \left(\frac{\partial}{\partial u} - \frac{1}{2}\frac{\partial}{\partial r}\right)b_{00'1} - \frac{\partial}{\partial r}b_{11'1} + \frac{1}{\sqrt{2}r}[(\mathscr{D}+\cot\theta)b_{10'1}$$

$$-(\bar{\mathscr{D}}+\cot\theta)b_{01'1}], \qquad \text{(C.16)}$$

$$(b+id)n_2 = \frac{1}{2r}(\mathscr{D}n_3\bar{\mathscr{D}}F - \bar{\mathscr{D}}n_3\mathscr{D}F) - \frac{1}{2}gF[(2b_{11'1} + b_{00'1})n_2$$

$$+ \sqrt{2}(b_{01'1}\bar{\mathscr{D}}n_2 - b_{10'1}\mathscr{D}n_2)], \qquad \text{(C.17)}$$

$$(b+id)n_3 = \frac{1}{2r}(\bar{\mathscr{D}}n_2\mathscr{D}F - \mathscr{D}n_2\bar{\mathscr{D}}F) - \frac{1}{2}gF[(2b_{11'1} + b_{00'1})n_3$$

$$+ \sqrt{2}(b_{01'1}\bar{\mathscr{D}}n_3 - b_{10'1}\mathscr{D}n_3)]. \qquad \text{(C.18)}$$

Using Eqs. (C.11), (C.12), (C.14) and (C.15) in Eqs. (C.5)–(C.8) one obtains, respectively

$$\frac{\partial}{\partial r}(b+id) = g(a+ic)Fn_1 + \frac{i}{F}(b+id)\left[(b-id)n_1 - i\frac{\partial}{\partial r}F\right], \qquad \text{(C.5a)}$$

$$\frac{1}{r}\bar{\mathscr{D}}(b+id) = g(a+ic)F\bar{\mathscr{D}}n_1 + \frac{i}{F}(b+id)\left[(b-id)\bar{\mathscr{D}}n_1 \frac{i}{r}\bar{\mathscr{D}}F\right], \quad (C.6a)$$

$$\frac{1}{r}\mathscr{D}(b+id) = g(a+ic)F\mathscr{D}n_1 + \frac{i}{F}(b+id)\left[(b-id)\mathscr{D}n_1 - \frac{i}{r}\mathscr{D}F\right], \quad (C.7a)$$

$$\left(\frac{\partial}{\partial u} - \frac{1}{2}\frac{\partial}{\partial r}\right)(b+id) = -\frac{1}{2}g(a+ic)Fn_1$$

$$-\frac{i}{2F}(b+id)\left[(b-id)n_1 + 2i\left(\frac{\partial}{\partial u} - \frac{1}{2}\frac{\partial}{\partial r}\right)F\right] \quad (C.8a)$$

The sum of one half of Eq. (C.5a) and Eq. (C.8a) gives

$$\frac{1}{(b+id)}\frac{\partial}{\partial u}(b+id) = \frac{1}{F}\frac{\partial}{\partial u}F. \quad (C.19)$$

Thus

$$(b+id) = fF, \quad (C.20)$$

where f is an arbitrary complex function of r, θ and ϕ only. Putting the last equation in (C.1) one gets

$$\frac{\partial}{\partial r}(a+ic) = 2gfF^2 n_1. \quad (C.1a)$$

But $a + ic$ is independent of u, then F is independent of u, and so $b + id$ is also independent of u.

From Eqs. (C.10)–(C.18) and arithmetical operations between them, one can get Eqs. (6.4.20)–(6.4.26) as follows:
Multiplying Eq. (C.15) by two and substract it from Eq. (C.11) and using the result that $\frac{\partial}{\partial u}F = 0$, one gets Eq. (6.4.20). Equation (6.4.21) is obtained if one multiplys Eq. (C.12) by $\mathscr{D}n_1$ and Eq. (C.14) by $2\bar{\mathscr{D}}n_1$, then subtracts the last from the first, using Eq. (6.4.20) and the fact that $n_1^2 + \bar{\mathscr{D}}n_1\mathscr{D}n_1 = 1$. Now multiplying Eq. (C.14) by two and adding it to the complex conjugate of (C.12),

one obtains Eq. (6.4.22). Also, multiplying Eq. (C.15) by two, subtracting it from the complex conjugate of Eq. (C.11), using Eq. (6.4.20) and the result that $\frac{\partial}{\partial u} F = 0$, one gets Eq. (6.4.23). Multiplying the complex conjugate of Eq. (C.13) by two, adding it to Eq. (C.10) and using $\frac{\partial}{\partial u} b_{01'1} = 0$, one obtains Eq. (6.4.24). Multiplying the complex conjugate of Eq. (C.13) by two, subtracting it from Eq. (C.10) and taking $\frac{\partial}{\partial u} b_{01'1} = 0$, one gets Eq. (6.4.25). The last equation (6.4.26) is obtained if one takes the real part of Eq. (C.16).

In the above we used $\frac{\partial}{\partial u} b_{01'1} = 0$, we will now show that all the potentials are independent of u. From the field equations (C.1)–(C.8) we get that a, b, c, d and F are independent of u. Then, the left side of Eq. (6.4.21) is independent of u. So the right side which is $b_{00'1} - 2b_{11'1}$ must be independent of u. If we use this result in Eq. (6.4.20) we find that also $b_{01'1} \bar{\mathscr{D}} n_1 + b_{10'1} \mathscr{D} n_1$ is independent of u. From Eq. (6.4.22) we find that $b_{00'1} + 2b_{11'1}$ is independent of u, so $b_{00'1}$ and $b_{11'1}$ are independent of u. From Eq. (6.4.23) we get that $b_{01'1} \bar{\mathscr{D}} n_1 - b_{10'1} \mathscr{D} n_1$ is independent of u, from this and $b_{01'1} \bar{\mathscr{D}} n_1 + b_{10'1} \mathscr{D} n_1$ that is independent of u one obtains also that $b_{01'1}$ and $b_{10'1}$ are independent of u. Since F is independent of u, all the other components of the potentials are independent of u.

Finally, Eqs. (6.4.35) and (6.4.36) are obtained by substituting the values of a, b, c, d, $b_{00'1}$ and $b_{01'1}$ from Eqs. (6.4.29)–(6.4.34) in Eqs. (C.1) and (C.5), using
$$\frac{\partial}{\partial r} = n_1 \frac{\partial}{\partial x}.$$

Appendix D: Derivation of Equations (6.4.37)–(6.4.44)

Here we use the solutions of the algebraic equations (6.4.8)–(6.4.11), Case 2 Eqs. (6.4.16)–(6.4.19) in the field equations and the relations between the potentials and the fields to get Eqs. (6.4.37)–(6.4.44). We start by inserting the values of the potentials from Eqs. (6.4.16)–(6.4.19) in Eqs. (B.18)–(B.29). This yields the following:

$$\frac{\partial}{\partial r}(a+ic) = -2ig(b+id)H, \tag{D.1}$$

$$\frac{1}{r}\bar{\mathscr{D}}(a+ic) = 0, \tag{D.2}$$

$$\frac{1}{r}\mathscr{D}(a+ic) = 0, \tag{D.3}$$

$$\left(\frac{\partial}{\partial u} - \frac{1}{2}\frac{\partial}{\partial r}\right)(a + ic) = -ig(b + id)H, \tag{D.4}$$

$$\frac{\partial}{\partial r}(b + id) = -ig[(a + ic)H + (b + id)(b_{00'1}n_1 + \sqrt{2}b_{10'1}\mathcal{D}n_1)], \tag{D.5}$$

$$\frac{1}{r}\bar{\mathcal{D}}(a + id) = ig(b + id)(\sqrt{2}b_{10'1}n_1 - b_{00'1}\bar{\mathcal{D}}n_1), \tag{D.6}$$

$$\frac{1}{2r}\mathcal{D}(b + id) = -ig(b + id)\left(b_{11'1}\mathcal{D}n_1 + \frac{1}{\sqrt{2}}b_{01'1}n_1\right), \tag{D.7}$$

$$\left(\frac{\partial}{\partial u} - \frac{1}{2}\frac{\partial}{\partial r}\right)(b + id) = ig\left[-\frac{1}{2}(a + ic)H + (b + id)\left(b_{11'1}n_1 - \frac{1}{\sqrt{2}}b_{01'1}\bar{\mathcal{D}}n_1\right)\right] \tag{D.8}$$

From Eqs. (D.2) and (D.3), one can show that $a + ic$ is independent of θ and ϕ.

Inserting the potentials from Eqs. (6.4.16)–(6.4.19) in the relations between the fields and the potentials Eqs. (2.2.23)–(2.2.25), then the components of those relations will be

$$[(a + ic) + igH^2]\mathcal{D}n_1 = \sqrt{2}\left(\frac{\partial}{\partial r} + \frac{1}{r}\right)b_{01'1} - \frac{1}{r}\mathcal{D}b_{00'1}, \tag{D.9}$$

$$\left[-(b + id) + \frac{\partial}{\partial r}H\right]\mathcal{D}n_2 - \frac{n_2}{r}\mathcal{D}H = gH(b_{00'1}\mathcal{D}n_3 - \sqrt{2}b_{01'1}n_3), \tag{D.10}$$

$$\left[-(b + id) + \frac{\partial}{\partial r}H\right]\mathcal{D}n_3 - \frac{n_3}{r}\mathcal{D}H = gH(\sqrt{2}b_{01'1}n_2 - b_{00'1}\mathcal{D}n_2), \tag{D.11}$$

$$[(a + ic) + igH^2]\bar{\mathcal{D}}n_1 = 2\sqrt{2}\left(\frac{\partial}{\partial u} - \frac{1}{2}\frac{\partial}{\partial r} - \frac{1}{2r}\right)b_{10'1} - \frac{2}{r}\bar{\mathcal{D}}b_{11'1}, \tag{D.12}$$

$$\left[b + id - 2\left(\frac{\partial}{\partial u} - \frac{1}{2}\frac{\partial}{\partial r}\right)H\right]\bar{\mathcal{D}}n_2 - \frac{n_2}{r}\bar{\mathcal{D}}H = -gH(\sqrt{2}b_{10'1}n_3 + 2b_{11'1}\bar{\mathcal{D}}n_3), \tag{D.13}$$

$$\left[b + id - 2\left(\frac{\partial}{\partial u} - \frac{1}{2}\frac{\partial}{\partial r}\right)H\right]\mathscr{D}n_3 - \frac{n_3}{r}\bar{\mathscr{D}}H = gH(\sqrt{2}b_{10'1}n_2 + 2b_{11'1}\bar{\mathscr{D}}n_2),$$

$$\text{(D.14)}$$

$$(a + ic)n_1 = \frac{\partial}{\partial u}b_{00'1} - i\left[gn_1H^2 + \frac{\sqrt{2}}{r}\text{Im}(\bar{\mathscr{D}} + \cot\theta)b_{01'1}\right], \quad \text{(D.15)}$$

$$(b + id)n_2 = n_2\frac{\partial}{\partial u}H + i\,\text{Im}\left(\frac{\mathscr{D}n_2}{r}\mathscr{D}H - \sqrt{2}gHb_{01'1}\bar{\mathscr{D}}n_3\right), \quad \text{(D.16)}$$

$$(b + id)n_3 = n_3\frac{\partial}{\partial u}H + i\,\text{Im}\left(\frac{\mathscr{D}n_3}{r}\mathscr{D}H - \sqrt{2}gHb_{10'1}\mathscr{D}n_2\right), \quad \text{(D.17)}$$

where Im denotes the imaginary part. In the last three equations, we used the fact that $b_{00'k} = -2b_{11'k}$ which is obvious for $k = 2, 3$ from Eqs. (6.4.16) and (6.4.17), for which $k = 1$ we will deduce this result from Eqs. (D.9)–(D.14) in the following.

Multiplying Eq. (D.10) by $\mathscr{D}n_3$ and Eq. (D.11) by $\mathscr{D}n_2$ and subtracting the last equation from the first yields

$$\frac{i}{r}\mathscr{D}H = gH(\sqrt{2}b_{01'1}n_1 - b_{00'1}\mathscr{D}n_1). \quad \text{(D.18)}$$

Inserting (D.18) in Eq. (D.10) or Eq. (D.11) yields

$$\frac{\partial}{\partial r}H - (b + id) = igH(b_{00'1}n_1 + \sqrt{2}b_{01'1}\bar{\mathscr{D}}n_1). \quad \text{(D.19)}$$

Multiplying Eq. (D.13) by $\bar{\mathscr{D}}n_3$ and Eq. (D.14) by $\bar{\mathscr{D}}n_2$ and subtracting the first from the last yields

$$\frac{i}{r}\bar{\mathscr{D}}H = -gH(\sqrt{2}b_{10'1}n_1 + 2b_{11'1}\bar{\mathscr{D}}n_1). \quad \text{(D.20)}$$

Putting (D.20) in Eq. (D.13) or Eq. (D.14), one gets

$$-2\left(\frac{\partial}{\partial u} - \frac{1}{2}\frac{\partial}{\partial r}\right)H + b + id = igH(-\sqrt{2}b_{10'1}\mathscr{D}n_1 + 2b_{11'1}n_1). \quad \text{(D.21)}$$

The sum of complex conjugate of Eq. (D.20) and Eq. (D.18) gives

$$b_{00'1} + 2b_{11'1} = 0, \tag{D.22}$$

that is, $b_{00'1} = -2b_{11'1}$, and the sum of Eqs. (D.19) and (D.21) gives

$$\left(\frac{\partial}{\partial u} - \frac{\partial}{\partial r}\right) H = \sqrt{2g H} \, \text{Im}(b_{01'1} \bar{\mathscr{D}} n_1). \tag{D.23}$$

From the real parts of Eqs. (D.15) and (D.16) one obtains, respectively

$$n_1 a = \frac{\partial}{\partial u} b_{00'1}, \tag{D.24}$$

$$b = \frac{\partial}{\partial u} H. \tag{D.25}$$

From the imaginary parts of the same equations one gets, respectively

$$n_1 c = -g n_1 H^2 - \frac{\sqrt{2}}{r} \text{Im}(\bar{\mathscr{D}} + \cot\theta) b_{01'1}, \tag{D.26}$$

$$n_2 d = \text{Im}\left(\frac{\bar{\mathscr{D}} n_2}{r} \mathscr{D} H - \sqrt{2g H} b_{01'1} \mathscr{D} n_3\right). \tag{D.27}$$

From the imaginary part of Eq. (D.17), one gets

$$n_3 d = \text{Im}\left(\frac{\bar{\mathscr{D}} n_3}{r} \mathscr{D} H - \sqrt{2g H} b_{10'1} \mathscr{D} n_2\right). \tag{D.28}$$

Now we use the results from the relations between the potentials and the fields in the field equations (D.1)–(D.8). From Eqs. (D.19) and (D.5) we get

$$\frac{\partial}{\partial r}(b + id) = -ig(a + ic)H + \frac{b + id}{H}\left[\frac{\partial}{\partial r} H - (b - id)\right]. \tag{D.29}$$

Equations (D.6) and (D.18) give

$$\frac{1}{b+id}\bar{\mathscr{D}}(b+id) = \frac{1}{H}\bar{\mathscr{D}}H, \tag{D.30}$$

and Eq. (D.7) with Eq. (D.20) yields

$$\frac{1}{b+id}\mathscr{D}(b+id) = \frac{1}{H}\mathscr{D}H. \tag{D.31}$$

Equations (D.8) and (D.21) give

$$\left(\frac{\partial}{\partial u}-\frac{1}{2}\frac{\partial}{\partial r}\right)(b+id) = -\frac{i}{2}g(a+ic)H - \frac{1}{2H}(b+id)\left[b-id - 2\left(\frac{\partial}{\partial u}-\frac{1}{2}\frac{\partial}{\partial r}\right)H\right] \tag{D.32}$$

From Eqs. (D.30) and (D.31) one gets

$$(b+id) = hH, \tag{D.33}$$

where h is an arbitrary complex function of u and r only. Putting the last result (D.33) in Eqs. (D.1) and (D.4) one gets

$$\frac{\partial}{\partial r}(a+ic) = -2ighH^2, \tag{D.34}$$

$$\left(\frac{\partial}{\partial u}-\frac{1}{2}\frac{\partial}{\partial r}\right)(a+ic) = -ighH^2. \tag{D.35}$$

But $a+ic$ and h are independent of θ and ϕ, thus H and $b+id$ are independent of θ and ϕ.

Since H is independent of θ and ϕ, then $\bar{\mathscr{D}}H = \mathscr{D}H = 0$. Using this in Eqs. (D.6) and (D.7), one obtains

$$\sqrt{2}b_{10'1}n_1 - b_{00'1}\bar{\mathscr{D}}n_1 = 0, \tag{D.36}$$

$$\sqrt{2}b_{01'1}n_1 + 2b_{11'1}\mathscr{D}n_1 = 0. \tag{D.37}$$

If we remember that $2b_{11'1} = -b_{00'1}$, we see that Eqs. (D.36) and (D.37) are the complex conjugate of one another and

$$b_{01'1} = \frac{1}{\sqrt{2n_1}} b_{00'1} \mathcal{D} n_1 . \tag{D.38}$$

Putting the last result (D.38) in Eq. (D.10) or Eq. (D.11) and using $\mathcal{D}H = 0$, one finds that

$$\frac{\partial}{\partial r} H - (b + id) = \frac{i}{n_1} gHb_{00'1} . \tag{D.39}$$

Then, from the real and the imaginary parts of the last equation one gets

$$\frac{\partial}{\partial r} H = b, \tag{D.40}$$

and

$$d = -\frac{1}{n_1} gHb_{00'1} . \tag{D.41}$$

But d and H are independent of θ and ϕ, so $\frac{1}{n_1} b_{00'1}$ must be independent of θ and ϕ and is given by

$$b_{00'1} = n_1 G , \tag{D.42}$$

where G is a real function of u and r only. From Eqs. (D.25) and (D.40) one gets

$$\frac{\partial}{\partial u} H = \frac{\partial}{\partial r} H , \tag{D.43}$$

thus H is a function of $u + r$ which equals t. Inserting Eqs. (D.38) and (D.42) in Eqs. (D.26) and (D.27), using $\mathcal{D}H = 0$, one obtains equations (6.4.41) and (6.4.42) respectively.

From Eqs. (D.1) and (D.4) one can show that $\frac{\partial}{\partial u}(a + ic) = \frac{\partial}{\partial r}(a + ic)$, so a

and c are functions of $u + r$. It can be shown from Eqs. (D.9), (D.38) and (D.42) that $a = \dfrac{\partial G}{\partial r}$. But from Eqs. (D.24) and (D.42), $a = \dfrac{\partial}{\partial u} G$, so

$$\frac{\partial G}{\partial u} = \frac{\partial G}{\partial r}, \tag{D.44}$$

and G is a function of $u + r = t$.

The differentiation with respect to u or r is identical to that with respect to t. Then from Eqs. (D.42), (D.22), (D.38), (D.25) and (D.24) one obtains Eqs. (6.4.37)–(6.4.40). Equations (6.4.43) and (6.4.44) are obtained by using the value of a, b, c and d, from Eqs. (6.4.39)–(6.4.42), in Eqs. (D.1) or (D.4) and (D.5) or (D.8), respectively.

References

1. C.N. Yang and R.L. Mills, *Phys. Rev.* **96** (1954) 191.
2. W. Heisenberg, *Z. Phys.* **77** (1932) 1.
3. E.P. Wigner, *Phys. Rev.* **51** (1937) 106.
4. B. Cassen and E.U. Condon, *Phys. Rev.* **50** (1936) 846.
5. W. Pauli, *Rev. Mod. Phys.* **13** (1941) 203.
6. R. Utiyama, *Phys. Rev.* **101** (1956) 1597.
7. J.C. Taylor, *Gauge Theories of Weak Interactions* (Cambridge University Press, Cambridge, 1976).
8. P.W. Higgs, *Phys. Lett.* **12** (1964) 132.
9. G. 't Hooft, *Nucl. Phys.* **B35** (1971) 167.
10. S. Weinberg, *Phys. Rev. Lett.* **19** (1967) 1264.
11. A. Salam, in *Elementary Particle Theory*, ed. N. Svartholm, Stockholm, 1968.
12. T.W.B. Kibble, *J. Math. Phys.* **2** (1961) 212.
13. M. Carmeli, *Group Theory and General Relativity* (McGraw-Hill, New York, 1977).
14. E.T. Newman and R. Penrose, *J. Math. Phys.* **3** (1962) 566.

15. M. Martellini and P. Sodano, *Phys. Rev.* **D22** (1980) 1325.
16. R. Jackiw, *Rev. Mod. Phys.* **49** (1977) 681.
17. R. Dashen, B. Hasslacher, and A. Neveu, *Phys. Rev.* **D10** (1974) 4114.
18. R. Dashen, B. Hasslacher, and A. Neveu, *Phys. Rev.* **D10** (1974) 4130.
19. V. Korepin and L. Faddeev, *Teor. Mat. Fiz.* **25** (1975) 147 [*Theor. Math. Phys.* **25** (1976) 1039].
20. J. Goldstone and R. Jackiw, *Phys. Rev.* **D11** (1975) 1486.
21. T.T. Wu and C.N. Yang, in *Properties of Matter under Unusual Conditions*, eds. H. Mark and S. Fernbach, (Interscience, New York, 1969).
22. H.G. Loos, *J. Math. Phys.* **8** (1967) 1870.
23. H.G. Loos, *J. Math. Phys.* **8** (1967) 2114.
24. R.P. Treat, *Nuovo Cimento* **50A** (1967) 871.
25. G. Rosen, *J. Math. Phys.* **13** (1972) 595.
26. H.B. Nielson and P. Oleson, *Nucl. Phys.* **B61** (1973) 45.
27. H. Georgi and S.L. Glashow, *Phys. Rev.* **D6** (1972) 2977.
28. G. 't Hooft, *Nucl. Phys.* **B79** (1974) 276.
29. A.M. Polyakov, *Zh. Eksp. Teor. Fiz. Pis'ma Red.* **20** (1974) 430 [*JETP-Lett.* **20** (1974) 194].
30. M.K. Prasad and C.M. Sommerfield, *Phys. Rev. Lett.* **35** (1975) 760.
31. J. Arafune, P.G.O. Freund, and C.J. Goebel, *J. Math. Phys.* **16** (1975) 433.
32. F.A. Bais and R.J. Russell, *Phys. Rev.* **D11** (1975) 2692.
33. B. Julia and A. Zee, *Phys. Rev.* **D11** (1975) 2227.
34. S. Deser, *Phys. Lett.* **64B** (1975) 463.
35. T. Eguchi, *Phys. Rev.* **D13** (1976) 1561.
36. P.H. Frampton, *Phys. Rev.* **D14** (1976) 528.
37. E.B. Bogomol'ny, *Sov. J. Nucl. Phys.* **24** (1976) 449.
38. E.B. Bogomol'ny and M.S. Marinov, *Sov. J. Nucl. Phys.* **23** (1976) 355.
39. J.P. Hsu, *Phys. Rev. Lett.* **36** (1976) 646.
40. Z.E.S. Uy, *Nucl. Phys.* **B110** (1976) 389.
41. E.J. Weinberg and A.H. Guth, *Phys. Rev.* **D14** (1976) 1660.
42. J.P. Hsu and E. Mac, *J. Math. Phys.* **18** (1977) 100.
43. G. 't Hooft, *Phys. Rev. Lett.* **37** (1976) 8.
44. G. 't Hooft, *Phys. Rev.* **D14** (1976) 3432.
45. E. Corrigan and D. Fairlie, *Phys. Lett.* **67B** (1977) 69.
46. F. Wilczek, in *Quark Confinement and Field Theory*, eds. D. Stump and D. Weingarten, (Wiley, New York, 1977).
47. A.P. Protogenov, *Phys. Lett.* **67B** (1977) 62.
48. L. O'Raifeartaigh, *Lett. Nuovo Cimento* **18** (1977) 205.
49. L. Michel, L.O'Raifeartaigh, and K.C. Wali, *Phys. Rev.* **D15** (1977) 3641.
50. R. Weder, *Commun. Math. Phys.* **57** (1977) 113.
51. S. Coleman, S. Parke, A. Neveu, and C.M. Sommerfield, *Phys. Rev.* **D15** (1977) 544.
52. M. Magg, *J. Math. Phys.* **19** (1978) 991.
53. M.A. Lohe, *Nucl. Phys.* **B142** (1978) 236.

54. S.F. Magruder, *Phys. Rev.* **D17** (1978) 3257.
55. C. Montonen and D. Olive, *Phys. Lett.* **72B** (1977) 117.
56. N.S. Manton, *Nucl. Phys.* **B135** (1978) 319.
57. E. Corrigan, D. Fairlie, S. Templeton, and P. Goddard, *Nucl. Phys.* **B140** (1978) 31.
58. C. Rebbi, *Phys. Rev.* **D17** (1978) 483.
59. M. Kalb, *Phys. Rev.* **D18** (1978) 2909.
60. M. Carmeli, *Phys. Lett.* **68B** (1977) 463.
61. S. Coleman, *Phys. Lett.* **70B** (1977) 59.
62. J.R. Morris, *Phys. Rev.* **D23** (1981) 556.
63. P. Goddard and D. Olive, *Rep. Prog. Phys.* **41** (1978) 1357.
64. R.S. Ward, *Commun. Math. Phys.* **79** (1981) 317.
65. M.K. Prasad, *Commun. Math. Phys.* **80** (1981) 137.
66. M.K. Prasad and P. Rossi, *Phys. Rev. Lett.* **46** (1981) 806.
67. A.A. Belavin, A.M. Polyakov, A.S. Schwartz, and Yu. S. Tyupkin, *Phys. Lett.* **59B** (1975) 85.
68. E. Witten, *Phys. Rev. Lett.* **38** (1977) 121.
69. M.F. Atiyah and R.S. Ward, *Commun. Math. Phys.* **55** (1977) 117.
70. V. de Alfaro, S. Fubini, and G. Furlan, *Phys. Lett.* **65B** (1976) 163.
71. V. de Alfaro, S. Fubini, and G. Furlan, *Phys. Lett.* **72B** (1977) 203.
72. J. Glimm and A. Jaffe, *Phys. Lett.* **73B** (1978) 167.
73. C. Callen, R. Dashen, and J. Gross, *Phys. Lett.* **66B** (1977) 375.
74. C. Callen, R. Dashen, and J. Gross, *Phys. Rev.* **D17** (1978) 2717.
75. A. Actor, *Rev. Mod. Phys.* **51** (1979) 461.
76. C.H. Oh, *Phys. Lett.* **74B** (1978) 239.
77. C.H. Oh and R. Teh, *Phys. Rev.* **D21** (1980) 531.
78. U. Sarkar and A. Raychaudhuri, *Phys. Rev.* **D26** (1982) 2804.
79. R. Teh, C.H. Oh, and W.K. Koo, *Phys. Rev.* **D26** (1982) 3649.
80. L. Mathelitsch, H. Mitter, and F. Widder, *Phys. Rev.* **D25** (1982) 1123.
81. C.H. Oh, *Phys. Rev.* **D25** (1982) 2194.
82. C.H. Oh, R. Teh, and W.K. Koo, *Phys. Rev.* **D25** (1982) 3263.
83. S.A. Brown, H. Panagopoulos, and M.K. Prasad, *Phys. Rev.* **D26** (1982) 854.
84. L. O'Raifeartaigh, S. Rouhani, and L.P. Singh, *Nucl. Phys.* **B206** (1982) 137.
85. E. Elizalde, *Phys. Rev.* **D27** (1983) 464.
86. S.A. Brown, *Phys. Rev.* **D27** (1983) 2968.
87. Shau-Jin Chang, *Phys. Rev.* **D29** (1984) 259.
88. M. Carmeli and Kh. Huleihil, *Nuovo Cimento* **67B** (1982) 21.
89. Kh. Huleihil, *Nuovo Cimento* **70A** (1982) 97.
90. M. Carmeli and Kh. Huleihil, *Nuovo Cimento* **74A** (1983) 245.
91. M. Carmeli and Kh. Huleihil, "Magnetic solution of Yang-Mills equations and the motion of classical particle," in *Differential Geometric Methods in Mathematical Physics*, eds. S. Sternberg, (D. Reidel Publishing Co., Dordrecht, Holland, 1984), pp. 145–159.
92. Kh. Huleihil, *Phys. Lett.* **138B** (1984) 287.

93. Kh. Huleihil, *Int. J. Theor. Phys.* **24** (1985) 571.
94. M. Carmeli, *Phys. Rev. Lett.* **39** (1977) 523.
95. M. Carmeli, *Phys. Lett.* **77B** (1978) 188.
96. M. Carmeli, in *Differential Geometrical Methods in Mathematical Physics*, eds. K. Bleuler, H.R. Petry, and A. Reetz (Springer-Verlag, Heidelberg, 1978).
97. M. Carmeli and D.H. Wohl, *Nuovo Cimento Lett.* **25** (1979) 230.
98. M. Carmeli and M. Fischler, *Phys. Rev.* **D19** (1979) 3653.
99. M. Carmeli, *Nuovo Cimento* **52A** (1979) 545.
100. M. Carmeli and B.Z. Moroz, in *Differential Geometrical Methods in Mathematical Physics*, eds. P.L. Garcia, A. Perez-Rendon, and J.M. Souriau (Springer-Verlag, New York, 1980).
101. L.L. Wang and C.N. Yang, *Phys. Rev.* **D17** (1978) 2687.
102. L. Castillejo, M. Kugler, and R.Z. Roskies, *Phys. Rev.* **D19** (1979) 1782.
103. J. Anandan and K.P. Tod, *Phys. Rev.* **D18** (1978) 1144.
104. L. Castillejo and M. Kugler, *Phys. Rev.* **D24** (1981) 2626.
105. W. Kinnersley, *J. Math. Phys.* **10** (1969) 1195.
106. M. Carmeli, Ch. Charach, and M. Kaye, *Nuovo Cimento* **45B** (1978) 310.
107. E.T. Newman, *Phys. Rev.* **D18** (1978) 2901.
108. E.T. Newman, *Phys. Rev.* **D22** (1980) 3023.
109. L. Altamirano and D. Villarroel, *Phys. Rev.* **D24** (1981) 3118.
110. S.K. Wong, *Nuovo Cimento* **65A** (1970) 689.
111. W. Drechsler, *Fortschr. Phys.* **27** (1979) 489.
112. W. Drechsler, *Phys. Lett.* **90B** (1980) 258.
113. W. Drechsler and A. Rosenblum, *Phys. Lett.* **106B** (1981) 81.
114. S. Ragusa, *Phys. Rev.* **D26** (1982) 1979.
115. C. Duval and P. Horváthy, *Ann. Phys.* **142** (1982) 10.
116. P. Horváthy, in *Proceedings of the IUTAM-ISIMN Symposium on Modern Developments in Analytical Mechanics*, Turin, June 7–11, 1982; *Att della Academia delle Scienze di Torino (Supp.)* **117** (1983) 163.
117. A.I. Janis and E.T. Newman, *J. Math. Phys.* **6** (1965) 902.
118. K. Wódkiewicz, *Acta Phys. Pol.* **6B** (1975) 509.
119. K. Wódkiewicz, *Phys. Rev.* **D11** (1975) 3395.
120. S. Malin, *Phys. Rev.* **D10** (1974) 2338.
121. M. Carmeli and M. Kaye, *Nuovo Cimento* **34B** (1976) 225.
122. M. Carmeli, *Classical Fields* (Wiley-Interscience, New York, 1982).
123. E.P. Wigner, *Group Theory*, New York, N.Y., 1959.
124. I.S. Gradshteyn and I.M. Ryzhik, *Table of Integrals, Series, and Products* (Academic Press, New York and London, 1965).
125. M. Mathisson, *Proc. Cambridge Philos. Soc.* **36** (1940) 331.
126. L.D. Landau and E.M. Lifshitz, *The Classical Theory of Fields* (Pergamon Press, New York, 1975), Sec. 22.

Subject Index

www.ingramcontent.com/pod-product-compliance
Lightning Source LLC
Chambersburg PA
CBHW050643190326
41458CB00008B/2396